The Biology of Chaetognaths

The Biology of Chaetognaths

Edited by

Q Bone
Marine Biological Association, Plymouth, UK

H Kapp
Zoologisches Institut u. Zoologisches Museum, Universität Hamburg and Biologische Anstalt Helgoland, Germany

and

A C Pierrot-Bults
Instituut v. Taxonomische Zoölogie, Universiteit v. Amsterdam, Nederland

OXFORD   NEW YORK   TOKYO
OXFORD UNIVERSITY PRESS
1991

Oxford University Press, Walton Street, Oxford OX2 6DP

Oxford New York Toronto
Delhi Bombay Calcutta Madras Karachi
Petaling Jaya Singapore Hong Kong Tokyo
Nairobi Dar es Salaam Cape Town
Melbourne Auckland
and associated companies in
Berlin Ibadan

Oxford is a trade mark of Oxford University Press

Published in the United States
by Oxford University Press, New York

A catalogue record for this book is available from the British Library

Library of Congress Cataloging in Publication Data
The Biology of chaetognaths / edited by Q. Bone, H. Kapp, A. C. Pierrot-Bults.
p.   cm.
1. Chaetognatha.   I. Bone, Q.   II. Kapp, H.   III. Pierrot-Bults, A. C.
QL 391.C6B56   1991   595.1′86—dc20   90-24016
ISBN 0–19–857715–X

Typeset by CentraCet, Cambridge
Printed in Great Britain by
The Bath Press, Avon

This book is dedicated to the distinguished Japanese worker on chaetognaths, Dr T. Tokioka of the Seto Marine Laboratory, in recognition of his contributions to the study of chaetognaths for over forty years.

CONTENTS

CONTRIBUTORS

*R. BIERI Antioch College, Yellow Springs, Ohio 45387, USA

Q. BONE Marine Biological Association of the United Kingdom, The Laboratory, Citadel Hill, Plymouth PL1 2PB, UK

D. V. P. CONWAY Plymouth Marine Laboratory, West Hoe, Plymouth PL1 3DH, UK

M. DUVERT Laboratoire de Cytologie, Université Bordeaux II, Avenue des Facultés, Talence-Cedex 33405, France

D. FEIGENBAUM 973 Edwin Drive, Virginia Beach, Virginia 23464, USA

T. GOTO Department of Biology, Faculty of Education, Mie University, Tsu-shi, Mie 514, Japan

H. KAPP Zoologisches Institut u. Zool. Museum, Universität Hamburg, Martin-Luther-King-Platz 3, 2000 Hamburg 13, Germany and Biologische Anstalt Helgoland, Notkestr. 31, 2000 Hamburg 52, Germany

S. NAGASAWA Ocean Research Institute, University of Tokyo 1-15-1 Minamidai, Nakano-ku, Tokyo 164, Japan

V. R. NAIR Regional Centre of the National Institute of Oceanography, Sea Shells. Seven Bungalows, Versova, Bombay 400 061, India

S. PEARRE Jr. Department of Oceanography, Dalhousie University, Halifax, Nova Scotia, Canada B3H 4J1

A. C. PIERROT-BULTS Institute of Taxonomic Zoology, University of Amsterdam, P.O. Box 4766, 1009 AT Amsterdam, The Netherlands

D. B. ROBINS Plymouth Marine Laboratory, West Hoe, Plymouth PL1 3DH, UK

M. TERAZAKI Ocean Research Institute, University of Tokyo 1-15-1 Minamidai, Nakano-ku, Tokyo 164, Japan

E. V. THUESEN Marine Science Institute and Department of Biological Sciences, University of California, Santa Barbara, California 93106, USA

*Dr Bieri died on 31 July, 1990.

1

INTRODUCTION AND RELATIONSHIPS OF THE GROUP

Q Bone, H Kapp, and A C Pierrot-Bults

Introduction

Chaetognaths are found in every marine habitat, from the benthos to all zones of coastal waters and the open oceans. Although small (2–120 mm long) they are often abundant, and play an important role in the marine food web as the primary predators of copepods. The biomass of chaetognaths has been estimated as 10–30 per cent of that of copepods in the world oceans, so that they are of great significance in transferring energy from copepods to higher trophic levels.

The first drawings of a chaetognath appeared in a paper by Slabber (1775), who called it a sea worm, arrow, or *Sagitta*; the second brief description was by Quoy and Gaimard (1827). None of these authors knew in which group to place the animals, and Leuckart and Pagenstecher (1858) proposed that the Chaetognatha should be classified as a separate group, to be placed between Annelida and Nematoda. Throughout the last century, many distinguished zoologists e.g. Darwin 1844; Krohn 1853; Hertwig 1880*a*,*b*,*c*; and Grassi 1883 examined chaetognath structure and embryology. A more complete list would include Huxley, Gegenbaur, Leydig, Metschnikoff, Kowalewsky, and Bütschli, and indeed would represent almost a roll-call of famous zoologists best-known for their work in other fields. Kuhl (1938) gives a detailed historical account.

The main aim of chaetognath workers in the last century was to determine the affinities of the group to other phyla. In this they were unsuccessful. Darwin spoke of the group as 'remarkable for the obscurity of its affinities', and Grassi, after nearly forty years of further studies, considered the problem of the affinities of chaetognaths as one of the most difficult that could be presented to a biologist, concluding (correctly!) that there was no sign that it would be solved soon.

Later work, while recognizing that the problem still remains, has concentrated on elucidating further details of morphology and behaviour, ecology, distribution, and taxonomy. Much of this more recent information about chaetognaths is scattered: the last comprehensive review of the group was published over fifty years ago by Kuhl (1938).

This book gives an up-to-date comprehensive review of different aspects of chaetognath biology, and is designed to be of practical value to plankton workers. The book begins with a brief account of chaetognath anatomy and morphology.

Chaetognaths consist of a head armed with grasping hooks and teeth and an elongated, fluid-filled body, which is a tube of locomotor musculature inside a tough basement membrane covered with a multi-layered epidermis, so that the musculature operates on a hydroskeleton. Apart from the gut and the gonads there are no internal organs. There are many unusual structural features, including a multi-layered epidermis, a retractile hood, and a 'corona ciliata' of unknown function. Whether or not the body cavity is a true coelom is still a matter of debate (Chapter 2 and 4). Since they are active predators, it is not surprizing that chaetognaths have a relatively complex nervous system (Chapter 3), with an elaborate array of vibration receptors which detect moving prey, and eyes of remarkably diverse structure in different species (some are even blind). Chaetognaths swim by dorsoventral oscillations of the body brought about by two distinct types of primary muscle fibres (Chapter 4), recalling those of arthropods in the structure of the contractile apparatus. They also have a secondary musculature whose fibres are unique in that the contractile apparatus consists of two quite different sarcomere types, alternating along the fibres. As yet, little is known of the roles of these different kinds of muscle fibres in the production of direction swimming movements. Some chaetognaths have special buoyancy devices which make them close to neutral buoyancy, but most are slightly more dense than sea-water, and swim in regular short bursts to maintain their position in the water column. Chaetognaths feed on a wide size range of prey (Chapter 5), the actual diet varying with prey abundance and season. Copepods (and sometimes small chaetognaths) are the main food items. The prey is captured with the hooks, and the much smaller teeth are perhaps used to puncture prey to aid in paralysing it before ingestion, although this is still conjectural. Paralysis is caused by the sodium channel blocker tetrodotoxin (TTX) (Chapter 6). This is produced by bacteria, but the exact location and manner of accumulation of bacterial TTX in the head is not yet known. Chaetognaths are protandric hermaphrodites (Chapter 7), and cross-fertilization seems to be almost universal (though this is still debated). Embryonic development differs from the typical deuterostome pattern, and development is continuous, without a metamorphosis. Several environmental factors such as temperature, salinity, oxygen, and food abundance influence the periodicity of spawning, growth and generation time. Bacterial infections (Chapter 8) are probably more fatal to chaetognaths than parasitic infections, although protozoan and nematode parasites are common and may produce behavioural and morphological changes. However, the effects of parasitism on population dynamics are still very poorly known. The distribution patterns of chaetognaths (Chapter 9) are essentially similar to those of other plankton and mikronekton groups, and suggest that they are an ancient oceanic group. Within the group, the distribution patterns suggest that the epipelagic genus *Pterosagitta* is more primitive than the epipelagic, mesopelagic, and bathypelagic genus *Sagitta*, and that the mesopelagic to bathypelagic *Eukrohnia* and the bathypelagic *Hetero-*

krohnia are more derived than *Sagitta*. The distribution patterns of deepsea chaetognaths (Chapter 10) are not well enough known to allow inferences about their relationships. Some deepsea chaetognaths brood their young in special marsupia, all others release the fertilized eggs into the sea. Chaetognath systematics were last revised at the beginning of this century (Ritter-Záhony 1911*a*,*b*). Since then a host of new species have been described, and the taxonomy of the group is still unclear. The division of the genus *Sagitta* was proposed by Tokioka (1965*a*) and although his scheme has not been widely accepted by later workers, it has provoked continuing debate. Most consider that Tokioka's subdivision leaves too many open questions about relationships, or have found it too complicated to use. In Chapter 10, an attempt is made to bring all chaetognath species described into a more comprehensive system of different genera. Although the arrangement given here is recognized to be tentative in several respects, and will certainly require future modification, the subdivision of the genus *Sagitta* aims to give more information about related species. Further work is required to decide whether the species groups recognized are really of generic status. Chaetognath structure presents special problems in collection and preservation (Chapter 11) to obtain dry weight and chemical composition values for energy flow studies. Culture methods are as yet only successful for more resilient neritic species. The experiments reported in Chapter 11 show the difficulties involved in obtaining accurate factors for estimates of chaetognath dry weight and chemical composition from preserved samples.

Throughout this book, outstanding problems and lines of future research are indicated, for much remains to be done in almost every aspect of chaetognath biology.

The affinities of the phylum

Since their discovery, chaetognaths have remained one of the most isolated phyla in the Animal Kingdom and although molecular techniques may resolve their true affinities, they are only beginning to be applied to the group. During the past 150 years, chaetognaths have been linked to an extraordinary list of other groups— Nematoda, Annelida, Mollusca, Crustacea, Arachnoidea, Protocoelomata, Oligomera, and even Chordata—sufficient indication that their true affinities are unknown. All the affinities suggested have already

been discussed by Grassi (1883), Kuhl (1938), Hyman (1959) and Ghirardelli (1968); the latter concluding with other more recent workers such as Ducret (1978) and Srinivasan (1979) that they are remote relatives of the Deuterostomata. Nevertheless, Casanova (1985*a*,*b*,*c*, 1986*a*,*b*,*c*,*d*,*e*, 1987) has revived the idea of a molluscan relationship, based on morphological similarities found between chaetognaths and a gymnosome opisthobranch. Rather than assisting in the elucidation of relationships with other groups, recent investigations of chaetognath fine structure have instead emphasized the isolation of the phylum (Duvert and Salat 1980; Duvert and Gros 1982; Bone *et al.* 1983; Bone and Pulsford 1984). Even where such studies have revealed similarities between chaetognaths and other groups at the ultrastructural level, the significance of these observations is far from clear, for example, chaetognath sperm have much in common with those of the arthropod-mollusc line (van Deurs 1972); cerebral ganglion ultrastructure resembles that of protostomes (Rehkämper and Welsch 1985); and the curious retrocerebral organ is very like that in some crustacea (see Chapter 5).

Since chaetognath embryonic development seems to be unique to the phylum, embryological investigations have not as yet provided any clues to affinities, but reinvestigation by modern techniques is badly needed to confirm and extend earlier studies. The features of ontogenesis which may be of phylogenetic significance are:

1. Cleavage is equal and radial.

2. The blastopore does not form the mouth, which appears at the opposite end of the embryo.

3. The mesoderm forms by the backward extension of a pair of folds, and the coelom does not form by enterocoely.

4. Development leads to a bipartite body, consisting of head and trunk. The tail is merely the posterior part of the trunk which is divided from the anterior in connection with the development of the gonads. Chaetognaths are not tripartite, as sometimes suggested.

Relationships within the Chaetognatha

The origin and evolutionary development of the group are unknown. *Amiskwia sagittiformis* Walcott, from the Burgess shale (Cambrian), has been supposed to be a chaetognath, although this has been refuted by Owre and Bayer (1962) and Conway-Morris (1977). Bieri (unpublished) has re-examined *Paucijaculum samamithion* (Schram 1973) from the Mazon Creek fauna (Upper Pennsylvanian), and has concluded that it is definitely a chaetognath (see Chapter 10). Apart from this species, and others in the Mazon Creek fauna, the only other possible chaetognath fossils are the small tooth-like protoconodonts, some of which show a striking resemblance to the grasping spines of modern chaetognaths (Szaniawski 1982). These occur from the base of the Cambrian to the lower Ordovician. It seems probable that the chaetognaths evolved at a time when the body-plans of most metazoans were established, around 550 million years ago (Conway-Morris 1987).

Studies of the structure and distribution of modern chaetognaths have not so far provided any agreed lines of evolution within the group. Several theories suggesting different basal genera have been proposed (e.g. *Spadella*: Tokioka 1965, 1966; Kassatkina 1980; *Eukrohnia*: Boltovskoy 1979; *Heterokrohnia*: Casanova 1987). No clear indication exists for the direction of evolution within the group, so it is not known whether the benthic *Spadella* gave rise to the pelagic forms or viceversa, and the possibility that the modern genera had a parallel rather than successive development cannot be excluded. Taking *Spadella* to be the basal genus, Tokioka (1965, 1966) suggested that the Pacific littoral was the site of origin of the group, but more recent knowledge of ocean-forming processes based on plate tectonics have resulted in this theory being discounted. Theories on the historical development of the ocean basins and the water masses within them have been developed by Van der Spoel (1983) and Van der Spoel and Heyman (1983). Hypotheses about distribution patterns and their historical development in the group (based on these theories) are developed in Chapter 9.

It has been known for some time that there are groups of species within the genus *Sagitta* that are more similar to each other than to other species or species groups, but, apart from the *serratodentata* group (and perhaps the *hexaptera* and *maxima* groups), such groupings have not been universally accepted (see Chapter 10). Similar attempts have been made for *Spadella* species (Tokioka and Pathansali 1964; Bieri 1974 *a,b*; Alvariño 1978, 1981; Salvini-Plawen 1986; Casanova 1987), and for *Eukrohnia* species (Casanova 1986). Tokioka (1965a) and Bieri (Chapter 10) divide the genus *Sagitta* into different genera to indicate related species, but the relationship between these genera remains unclear. The rationale for subdividing the genus *Sagitta* is discussed in Chapter 10; here it will suffice to say that any genus containing species so different as *S. elegans* and *S. setosa*, for example, can hardly be satisfactory. Interesting preliminary attempts have been made to use numerical methods to examine the relationships between chaetognath genera and species groups within genera (Dallot and Ibanez 1972; Salvini-Plawen 1986), these may well prove more successful as more is discovered about chaetognath structure. A more promising approach is certainly that of examining sequence similarities in the variable regions of the 18 and 28S rDNA loci of different species. Application of molecular techniques of this kind should provide an acceptable arrangement of the chaetognath genera and species groups in the next few years.

2

MORPHOLOGY AND ANATOMY
H Kapp

Introduction and general morphology

Chaetognaths are elongated, bilaterally symmetrical marine animals; they may be planktonic or benthic. Their body, which is more or less circular in section, bears one or two pairs of lateral fins and a tail fin. Chaetognaths are named from their large cephalic grasping spines ($\chi\alpha\iota\varsigma$ = chaete = spine, $\gamma\nu\alpha\tau o\varsigma$ = gnathos = jaw). Their common name, arrow worms, aptly describes both their body form (Fig. 2.1) and their remarkably rapid darting movements. The majority of species are almost completely transparent, but some are translucent and others contain blue, brown, orange, or red pigments. Overall length ranges from 2–120 mm. A rounded and somewhat flattened head bearing grasping spines or hooks, teeth, vestibular organs, and a ventral mouth (Fig. 2.2) is separated by a transverse septum from a long muscular trunk. The gut extends through the trunk, terminating at a ventral anus just anterior to a second transverse septum, which divides the posterior trunk region from the tail. Paired female reproductive organs lie in the trunk; the male reproductive system is situated in and on the tail. Conspicuous ciliary fan, or fence, receptor organs are seen arrayed on the head, trunk, tail and even fins of living specimens. On the posterior dorsal surface of the head, behind the eyes, there is a ciliary loop, or 'corona ciliata', which extends posteriorly onto the trunk in many species.

The peculiar morphology and anatomy of the group has been recognized ever since the earliest studies (e.g. Darwin, 1844), which placed chaetognaths in an isolated position with respect to all other invertebrate phyla. Hertwig (1880*a,c*) and Grassi (1883) confirmed the isolation of the group and established much of our knowledge of chaetognath anatomy; see the comprehensive treatises of Kuhl (1938), Hyman (1959), and Ghirardelli (1968). More recently, many workers have provided new information using modern techniques, especially electron microscopy, and many new and interesting species have been described. However, despite this spate of recent work, chaetognath anatomy is still relatively poorly known in some respects, and little or nothing is known of the function of many structures. Hence several features of anatomy and

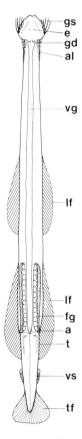

Fig. 2.1 Schematic diagram of chaetognath (*Sagitta*). a, anus; al, alveolar tissue; e, eye; fg, female gonad; gd, gut diverticle; gs, grasping spines; lf, lateral fin; t, testis; tf, tail fin; vg, ventral ganglion; vs, seminal vesicle.

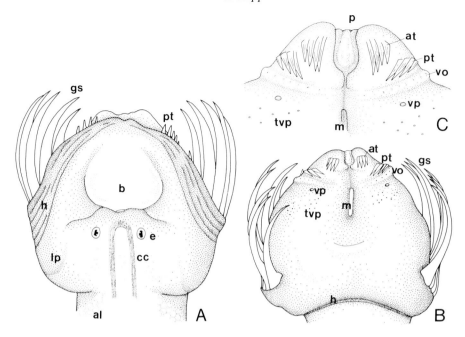

Fig. 2.2   (A) Head, dorsal view, (B) head, ventral view, (C) anterior part of head, ventral view. al, alveolar tissue; at, anterior teeth; b, brain; cc, corona ciliata; e, eye; gs, grasping spines; h, hood; lp, lateral plate; m, mouth; p, apical pit; pt, posterior teeth; tvp, transvestibular pores; vo, vestibular organ; vp, vestibular pit.

embryonic development which have already been investigated at the light microscopic level require reinvestigation, and experimental approaches are needed to determine the function of many of the more curious structures. Furthermore, the majority of the detailed studies have been carried out on only few species of *Sagitta* or *Spadella*—more species and genera deserve investigation.

The integument and its derivatives

Epidermis

The epidermis over the trunk and dorsal surface of the head is multilayered, unusually for an invertebrate (Furnestin 1967; Moreno 1975; Duvert and Salat 1979; Ahnelt 1980, 1984; Welsch and Storch 1983*b*, Duvert *et al*. 1984; Bouligand 1985). Flattened polygonal cells form the epidermal surface. These cells contain basal and lateral layers of tonofilaments, or intermediate filaments, and are covered by a thin network of filaments (0·4 μm in diameter), which are rich in anionic groups and are secreted by the outer zones of the surface cells. Additional secretory cytoplasmatic vesicles are secreted, to stick on the body surface. Below the surface cells there are two or more layers of compact cells containing abundant bundles of tonofilaments. which extend into the evaginations of the cell membrane (Fig. 2.3(A)). These interlink with deep invaginations of the neighbouring cells; and thus the coherence of the epidermis results from a three-dimensional system of interdigitations. The compact cells are flattened over thick nerves and the ventral ganglion. Duvert (1979), Welsch (1983*b*) and their co-authors distinguish two types of compact cells: those with abundant bundles of tonofilaments in the deepest layer, and those with fewer microfilaments between the deepest layer and the surface cells.

The integument is separated from the underlying tissue, mainly musculature, by a continuous basement membrane consisting of the basal laminae of the epi-

Fig. 2.3 (A) Compact cell of integument (*Sagitta elegans*), (B) two views of basement membrane (*S. elegans*), (C) vestibular papillae (*S. hexaptera*), (D) dentated hooks (*Heterokrohnia longidentata*), (E) teeth with cusps at the tips (*S. enflata*). af, attached filaments; k, collagenous fibres; m, mitochondrion, r, radial section of attachment zone; t, tangential section of attachment zone; w, wavy pattern of tonofilaments. Photographs by courtesy of P K Ahnelt (A) and (B), E V Thuesen (C), and I Moreno (E).

dermis and muscle cells, and a layer (0.2–0.3 μm thick) interposed between them. This layer contains a matrix and collagenous fibres which are obliquely orientated in alternating directions (at approximately 50° to the body axis) and arranged in sublayers, which are also orientated alternately, comparable to a fibre-wound cable (Fig. 2.4). Thus, the basement membrane and the compact epidermal cells form a sort of exoskeleton, giving both firmness and flexibility. The basement membrane prevents all physical contact between the epidermis and musculature but is not a barrier to diffusion; e.g. it is permeable to ions and acetylcholine.

Duvert (personal communication) regards a hydroskeleton as an important element in the construction of chaetognaths. This hydroskeleton consists of two main components: the basement membrane together with the compact cells, and the fluid of the body cavities. There may be a difference in the hydroskeleton between species, with vacuolized gut cells filling the trunk cavity (see page 14) and those without compartments in the trunk cavity.

The alveolar tissue (collarette) consists of specially differentiated epidermal cells. Each cell has a large vacuole surrounded by cytoplasm and a layer of tonofilaments. The tissue is found over part or all the body, depending on the species. In some it occurs on the neck, in others on parts of the trunk and tail, and in juveniles of *Eukrohnia*, in *Heterokrohnia involucrum* and in *H. furnestinae* it covers the entire body surface. The function of this tissue remains to be determined, but it may possibly assist buoyancy or serve as a protective layer.

A monolayered epidermis is found on the ventral surface of the head, on the fins, eyes, and the inner surface of the hood. The epidermis found on the ventral surface of the head consists of tall columnar cells which produce a thick cuticle (Fig. 2.5). The cuticle appears to be of an amorphous material with a certain degree of heterogeneity in electron micrographs, its surface is covered by a thin filamentous coat. Duvert and Salat (1979) found the cuticle present only on the head, but Furnestin (1967) published a micrograph showing a portion of tail epidermis of *S. elegans* covered by cuticle and Pierrot-Bults (1976*b*) observed a thin cuticle over the seminal vesicles of the *S. serratodentata* group.

Grasping spines and teeth

The grasping spines are much larger and longer than the teeth but are of essentially similar structure. They are constructed of two concentric tubes of α-chitin, connected by obliquely arranged lamellae (which look like oblique fibres under the light microscope), and by a flange along the inner curve (Fig. 2.6). The chitinous tubes contain zinc and the teeth and spines are tipped with silicon, which presumably hardens them. The tips of the spines are fluted and appear to consist of a series of elongate processes held together in a fibrous sheath. Cells around the bases of the spines and teeth send their processes into the 'pulp cavity' (Fig. 2.6); these processes contain axially-orientated microtubules and vesicles and mitochondria. Near the bases, where the teeth and spines are secreted, vesicles within the basal cells contain zinc, suggesting that the zinc within the spines and teeth is derived from the diet rather than absorbed directly from the sea water. The teeth and hooks lie in relatively flexible cuticular pockets and are surrounded and underlain by connective tissue 'anchor' cells, which send processes into the cuticular pockets and link teeth and spines to the muscle bundles controlling their movements (Bone *et al.* 1983).

The number and form of grasping spines and teeth vary according to the species. The number increases up

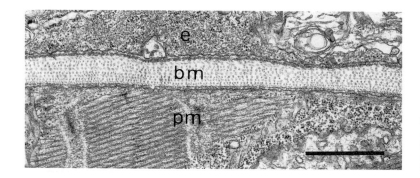

Fig. 2.4 Basement membrane of *S. setosa* showing layers of collagen fibres. bm, basement membrane; e, epidermis; pm, primary muscles. Scale bar: 1 μm. Photograph by M Duvert.

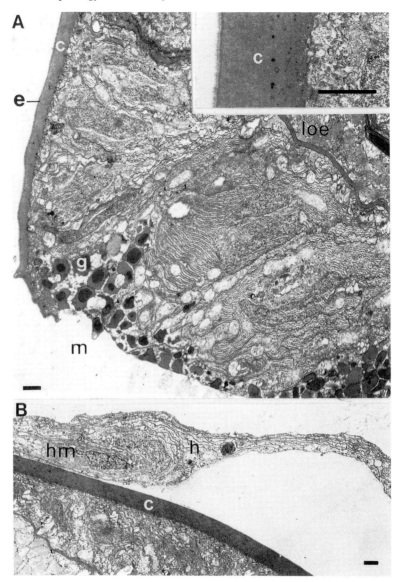

Fig. 2.5 (A) Cuticle of ventral surface of head in *S. setosa*, inset detail. c, cuticle; e, epidermis; g, glandular epithelium; m, mouth; loe, longitudinal oesophageal muscle. Scale bars: 1 μm. (B) Section of hood lying against head. c, cuticle; h, hood; hm, hood muscle. Scale bar: 1 μm. Photographs by M Duvert.

to maturity and then decreases; some are lost in older animals. Juveniles of *Eukrohnia* and *Heterokrohnia* have a set of dentated hooks which are lost with age (Furnestin 1965, Kapp and Hagen 1985) (see Fig. 2.3(D)). A group of *Sagitta* species possess serrated grasping spines in all stages of their ontogeny. Slight differences are seen in the shape of the hooks (Nagasawa and Marumo 1973), with the tips following the

spine curve or being bent. In addition, electron micrographs of the teeth have revealed a great number of different, species-specific, surface structures, especially of the anterior teeth (Cosper and Reeve 1970; Furnestin 1977, 1982; Moreno 1979; Kapp and Hagen 1985) and have shown the tips of the teeth in some species to be formed by several cusps (Bieri *et al.* 1983; Thuesen and Bieri 1987) (Fig. 2.3(E)).

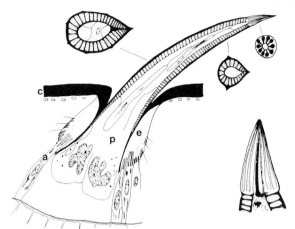

Fig. 2.6 Schematic diagram of a grasping spine. a, anchor cell; c, cuticular sheet; e, pocket of electron-lucent chitin; p, cells of pulp cavity. Redrawn from Bone *et al.* (1983).

Lateral and ventral plates

These are chitinous structures lying on the head epidermis and covered by the cuticle. They have been shown well in section by Burfield (1927). The lateral plates run along almost the entire laterodorsal length of the head and the ventral plates are approximately triangular and lie lateroventrally (Kuhl 1938). Both pairs of plates serve as skeletal attachments for the complex head musculature.

Vestibular organs, vestibular pits and other pores

The toxic and sticky secretions which are used to entangle and paralyse the prey (Chapter 5 and 6) are produced in the upper part of the oesophagus and in the vestibular organs and pits (Parry 1944; Bieri *et al.* 1983; Thuesen and Bieri 1987; Thuesen *et al.* 1988a; Kapp and Mathey 1989). The vestibular organs are oblique transverse papillated ridges lying just behind the posterior teeth; in *Heterokrohnia* irregular forms also occur. The papillae vary in form and size according to the species. Thuesen (1987, 1988a,b), and Kapp (1989) and their co-authors observed multipored openings at the tips of the papillae (see Fig. 2.3(C)); Bone and Pulsford (1984) and Ahnelt (1980) found cilia of sensory cells (Chapter 3) at the papillae. In section, these cilia are seen to have multiple tubules, they arise from deep within the sensory cells which lie amongst elongate glandular cells.

The vestibular pits lie posterior to the vestibular ridges and, although scarcely visible under the light microscope, are obvious in scanning micrographs. As Burfield (1927) and Thuesen and Bieri (1987) have shown, the pits are the openings of small organs consisting of elongate columnar secretory cells.

Posterior to the vestibular pits, on the ventral cuticle shielding the ventral surface of the head, there are scattered pores and cilia—the transvestibular pores and transvestibular cilia—which differ in number and position according to species (Ahnelt 1980, 1984; Bone and Pulsford 1984; Thuesen *et al.* 1988b). These pores are the openings of further ciliated sensory cells. Finally, intrabuccal pores have been described in *S. hispida* (Thuesen *et al.* 1988b) and a pair of narial pores have been found between the anterior teeth in *S. hexaptera* (Thuesen and Bieri 1987). Experimental evidence for the function of this wide array of sensory cells is lacking, but it seems highly probable that most are chemosensory.

Apical glandular tissue, neck channels, tentacles, and palps

Conspicuous cell complexes cap the front of the head in *Eukrohnia* and *Heterokrohnia* species. Their elongated cells seem to be glandular but, like the tube-like neck channels which reach from the head to the anterior trunk, they have not been studied in detail. A pair of delicate tentacles with more or less numerous papillae is situated near the base of the grasping spines or between the anterior teeth and the mouth in some *Spadella* species (John 1933; Owre 1963). In *Heterokrohnia palpifera*, regarded as a benthopelagic form, there is a pair of well developed, denticulate or papillate palps behind the vestibular organs (Casanova 1986d; Kapp, in press). Nothing is known about the function of these tentacles and palps.

Hood or praeputium

The hood (unique to chaetognaths) is an extension of the epidermis of the head; it is provided with retractor and extensor muscles. When extended, it enwraps the head completely, except for a round hole above the mouth, streamlining the head. It can also be retracted rapidly, to permit the grasping spines to spread out unhindered. Ventrally, the hood is attached at the border between head and trunk; laterally the attachment curves into an inverted V on the dorsal surface of the head. The outer wall of the hood is composed of normal epidermal cells; the inner wall is a glandular epithelium (Kuhl 1938). Glandular cells are also found on the surface of the head, they are arranged in a line at the attachment of the hood (Ritter-Záhony 1911*b*). The glandular secretions of both of these types of cell permit the hood to retract easily over the head surface,—this is vital when the grasping spines are being rapidly spread.

Corona ciliata or ciliary loop

In most species the corona ciliata is a more or less elongate oval, with a regular or wavy edge; in preserved specimens it is often damaged or lost. Beginning anteriorly behind the retrocerebral opening, or behind the eyes, it lies only on the head, on the neck or, according to species, extends a greater or lesser distance along the trunk. In pelagic species, the corona ciliata consists of two concentric rows of ciliated cells on the epidermal surface, bearing short (approximately 10 μm) cilia—much shorter than those of the ciliary fence receptors (Spero *et al.* 1979). In *Spadella*, the corona ciliata is elliptical and lies across the neck, where an inner ring of secretory cells is surrounded by a single ring of ciliated cells. Both types of cell arise from the same embryonal cells.

The function of the ciliary loop is unclear, although there is some evidence that it may be sensory (Chapter 3). In *Spadella* species it may play a role in reproduction, since secretion of the inner glandular cells is distributed along the medial dorsal trunk surface, leading to the gonopores and, as Ghirardelli (1968) pointed out, 'this distribution coincides remarkably with the course followed by spermatozoa after copulation.'

Adhesive organs of *Spadella* species

Juveniles of *Spadella schizoptera* can attach to the substrate by adhesive organs which consist of numerous papillae surrounding the anterior part of the head (Feigenbaum 1976). Later in development the specimens adhere by posterior adhesive organs. Paired epidermal digitations are found on the posterior end of the lateral fins, on the anterior end of the caudal fin or on the ventrolateral sides of the posterior part of the tail segment (Mawson 1944; Owre 1963, 1972; Alvariño 1970, 1981, 1987). These digitations contain muscle fibres derived from both dorsal and ventral muscle bands and their tips are covered by tubercles or papillae (Fig. 2.7(A) and (B)). In *S. cephaloptera* there are clusters of adhesive papillae on the ventral surface of the tail; these clusters are more numerous in the

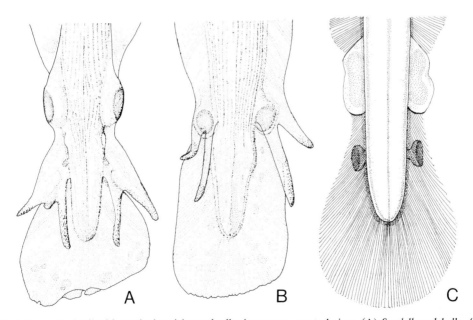

Fig. 2.7 Posterior part of tail with seminal vesicles and adhesive organs, ventral view. (A) *Spadella pulchella*, (B) *S. nana*, redrawn from H B Owre (1963), (C) *S. gaetanoi*, dorsal view with cup-like structures. Redrawn and modified from A Alvariño (1978).

anterior tail. According to John (1933) they function in a similar way to the tube feet of echinoderms. The pair of cup-like structures on the tail of *S. gaetanoi* described by Alvariño (1978) also may have an adhesive function (Fig. 2.7 (C)).

Fins

The fins are covered by a thin, monolayered epidermis which is stiffened by two rows of chitinous rays and filled with an amorphous extracellular matrix (Welsch and Storch 1983*b*, Ahnelt 1984). Ciliary fence receptors are often present on the fin surfaces. Basally, the fins are more or less wedge-shaped and in several species, as the animals mature (or possibly earlier), the extra-cellular matrix forming the wedges is expanded into more or less large gelatinous masses, perhaps to com-pensate for the increasing weight of gonads or to assist buoyancy. Fin shapes differ in different genera and species (see Chapter 11), but not enough is known about the swimming behaviour of chaetognaths to relate fin shape and area to possible differences between species in locomotor behaviour.

For the nervous system, retrocerebral organ, and sense organs see Chapter 3; see Chapter 4 for muscles.

Body cavity

The existence of a 'true' coelom in adult chaetognaths has long been debated (Hyman 1959; Ahnelt 1980). Ultrastructural studies of *Sagitta* species (Duvert and Salat 1979; Welsch and Storch 1982) have shown that the fluid-filled body cavity is lined with a thin epithe-lium. Welsch and Storch (1982) considered this to be of mesodermal origin and, even if the original meso-dermal lumen disappeared during development (see p. 16), felt it justifiable to term the body cavity a coelom. However, Duvert (personal communication) has come to the opposite conclusion. Reinvestigation of the embryonic development by modern techniques is required to settle the status of the body cavity.

The epithelium covering the muscle layer is extremely thin and is continuous with that overlying the mesenteries which attach the gut to the body wall and with the epithelium overlying the gut. Welsch and Storch (1982) found that the epithelium covering the gut in *Sagitta setosa* is thicker than that over the musculature and that it contains myofilaments sur-rounding the intestine. Also in *S. setosa* Bone *et al.* (1987*a*) described unstriated muscle cells surrounding the gut (Chapter 4) overlain by a possibly interrupted layer of thin epithelial cells. The thin epithelium sep-arating the fluid within the body cavity from the surrounding tissues may play an important role in controlling the pressure and chemical composition of the body fluid. At the lateral fields, where musculature is absent, the epithelium is thicker and has a different structure—the prismatic cells are rich in glycogen and granular reticulum and appear to be secretory, possibly contributing to the fluid within the body cavity. Some of these cells are ciliated, but the function of the cilia is unclear.

Digestive organs

The complete digestive system consists of a continuous tube, subdivided histologically and functionally into separate sections: mouth, oesophagus, intestine, rectum, and anus.

Mouth

The very extensible slit-like mouth opening, with its long dimension parallel to the body axis, lies in the centre of the vestibulum—a depression in the median ventral region of the head. The epidermis of the oral slit is more flattened than that of the vestibulum and is provided with numerous sensory cells (Bone and Puls-ford 1984). It is continuous with the secretory epithe-lium of the oesophagus.

Oesophagus

Initially relatively narrow, the oesophagus, or pharynx, expands to a bulb and narrows again before joining the intestine in the neck region. It is surrounded by circular muscles and, laterally, by small bundles of longitudinal

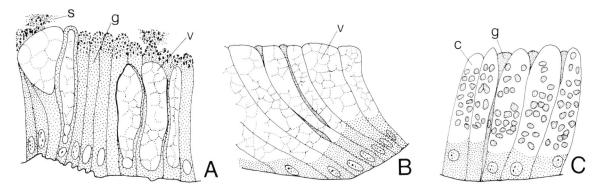

Fig. 2.8 Cell types of oesophagus. (A) *Spadella cephaloptera*, (B) and (C) *Sagitta setosa*. c, compound granular cells; g, granular cells; s, free secretion; v, vacuolated cells. Redrawn and modified from Parry (1944).

muscles (John 1933; Parry 1944). Parry (1944) and Stovall, (in Feigenbaum and Maris 1984) distinguished different types of cells in the monolayered epithelium of the oesophagus: granular and vacuolated cells in *Spadella*; granular, vacuolated and compound granular cells in *Sagitta*. The granular cells are columnar and distally filled with a granular secretion (secretion granules are also present in the lumen of the pharynx, embedded in a matrix). The vacuolated cells contain a large distal vacuole, which is sometimes subdivided into faint, spherical compartments. The compound granular cells are tall columnar cells with large inclusions containing granules (Fig. 2.8). All authors have suggested that the epithelium of the oesophagus produces a variety of secretions, including a thin peritrophic membrane.

Intestine

The intestine has no appendages except the gut diverticula, lateral and anteriorly-directed protrusions which occur only in some species. It is attached to the body walls dorsally and ventrally by longitudinal mesenteries and surrounded by a fine endothelium (see p. 12). The cells of the monolayered intestine epithelium interlock by evaginations and invaginations of their lateral walls and are linked distally by septate junctions and basally by gap junctions (Duvert *et al.* 1980a; Welsch and Storch 1983a). Two types of cells, glandular and absorptive, were found by Welsch and Storch (1983a) and by Parry (1944). Parry observed considerably more secretory cells in the anterior part of the intestine and considerably more absorptive cells in the posterior region. The glandular cells contain vacuoles filled with

secretions. After feeding they appear small and disrupted and it is some hours before they once more contain large vacuoles of secretions. The absorptive cells appear to be finely granulated and are ciliated. They show large vesicles some time after feeding, which slowly diminish as their contents are apparently absorbed by the intestinal cells (Fig. 2.9). Nothing is

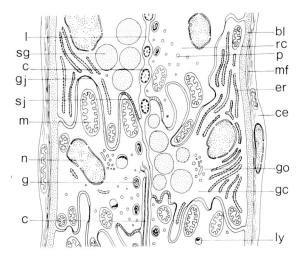

Fig. 2.9 Schematic diagram of intestine, redrawn and modified from Welsch and Storch (1983a). bl, basal lamina; c, cilia; ce, coelom epithelium; er, rough ER cisternae; g, apical secretion granules; gc, gland cell; gj, gap junction ; go, golgi complex; l, lumen; ly, lysosomal vesicles; m, mitochondrion; mf, muscle fibres; n, nucleus; p, (pinosomal?) vesicles; rc, resorptive cell; sg, secretion granules; sj, septate junction.

known of the manner whereby nutrients pass from the absorptive cells to the other tissues.

The vacuolized cells in the side walls of the gut were first detected by Meek (1928) in *Sagitta elegans* and described by Dallot (1970) in several *Sagitta* species. They contain large sac-like vacuoles and differ in size and number according to species, for example, in *S. elegans* (Bone *et al.* 1987*a*) they almost fill the whole trunk cavity and they contain NH_4 as a buoyancy device (Chapter 4).

Rectum and anus

The short rectum consists of regular, columnar ciliated cells and is surrounded by circular muscles. Its epithelium and the partly thickened basement membrane are extremely extensible. Remarkably, the anus lacks a circular sphincter muscle.

Reproductive organs

Chaetognaths are protandric hermaphrodites. The male and female reproductive organs are completely separated by the septum between trunk and tail. There are some unusual features in their organization, which is unlike that found in any other phylum.

Female reproductive organs

The female reproductive organs are more or less elongate cylinders lying in the posterior part of the trunk between the intestine and the body wall. Their size is species-specific and depends on the stage of maturity. Thin mesenteries continuous with the thin, covering endothelium attach the female organs laterally to the body wall. Anteriorly, each mesentery forms a triangular lamina. In some species a ligament (Grassi 1883) connects the inner posterior end of the ovary and the

mid-ventral region of the trunk just above the transverse septum, in some other species there are ducts (e.g. Mawson 1944; Casanova and Chidgey 1987). The larger part of the inner region of the female organs is filled with oocytes at different stages of development (Ghirardelli 1968). On its outer side the seminal receptacle, or ovispermaduct, runs along the entire length of the ovary. It is two-layered, consisting of an inner syncytial layer and an outer tube of cuboid or columnar cells. The shape of this outer tube differs between species, being in some a simple cell layer with longer lateral than medial cells (e.g. *Sagitta planctonis*, Pierrot-Bults 1975*b*), whereas in others its cross-section is crescentic with the oocytes in its concavity (e.g. *Sagitta bipunctata*, Ghirardelli 1968; Fig. 2.10). The duct curves almost at right angles to reach the external gonopore, opening dorsolaterally on a small papilla. In some

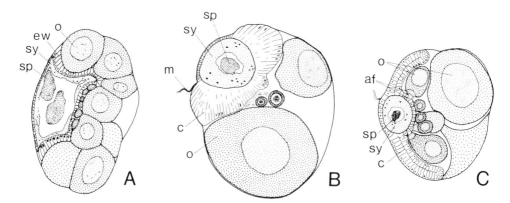

Fig. 2.10 Cross sections of ovaries. (A) *Sagitta planctonis*, (B) *S. enflata*, (C) *S. bipunctata*. af, accessory fertilization cells; c, crescent; ew, epithelial wall; m, mesentery; o, oocyte; sp, sperm; sy, syncytium. Drawn from photographs taken by A C Pierrot Bults (1975*b*) (A) and E Ghirardelli (1968) (B and C).

species, the terminal end of the duct is lined by a glandular epithelium secreting gelatinous material in which the eggs are embedded (*Pterosagitta draco*) or attached to the substrate (*Spadella*). Sometimes, part of the duct is enlarged as a spherical or club-shaped ampulla serving as a sperm reservoir; this ampulla lies either at the anterior end or near the posterior end of the duct. The oocytes are linked to the duct by a stalk formed by two suspension or accessory fertilization cells. The posterior cell is elongate and the other, touching the oocyte, is flattened and reniform. The vitelline membrane of the oocyte is interrupted where it is in contact with the suspension cell; the basement membranes of the suspension cells are also discontinuous and there is a series of vacuoles in their cytoplasm. These cells form the pathway by which sperm migrates from the seminal receptacle to penetrate the oocyte. Some authors have suggested that the suspension cells may have a trophic function for the oocytes (Pierrot-Bults 1975*b*). After fertilization, the suspension cells degenerate and the ova pass to the exterior. Although previous accounts suggested that a temporary oviduct is formed between the syncytium and the cellular wall of the seminal receptacle (Ghirardelli 1968), more recent investigations have revealed that the eggs reach the exterior via the ovispermaduct (*Sagitta hispida*, Reeve and Lester 1974; *S. planctonis*, Pierrot-Bults 1975*b*).

Some *Eukrohnia* species possess marsupial sacs and brood their young (see Chapter 10).

Male reproductive organs

The testes lie posterior to the trunk–tail septum and are surrounded by a thin endothelium which is continuous with a thin mesentery attaching them to the lateral walls of the tail. Clusters of spermatogonia freed from the testes mature to spermatocytes and spermatozoa whilst floating in the tail cavity, which is completely divided by a median septum and further subdivided by two incomplete secondary longitudinal septa. The walls of these cavities are ciliated and the sperm or spermatocyte morulae are thus circulated continually, upwards near the body wall, downwards near the middle septum; some species lack the secondary septa.

Mature sperm passes through the vasa deferentia into the seminal vesicles. The openings of the vasa deferentia are funnel-shaped and ciliated, their more or less elongated ducts run along the lateral walls of the tail, which they pierce to end in the anterior part of the seminal vesicles.

The seminal vesicles have a species-specific structure, ranging from a simple oval epidermal pouch lined with a glandular epithelium to a much more complex organization. In some *Sagitta* species, for example, they are elongate with a terminal knob filled with glandular secretion, and in the *Sagitta serratodentata* group (Pierrot-Bults 1974, 1976*b*) they are not only bipartite, but provided with external structures. There seems to be a preformed site of initial rupture in these more complex seminal vesicles which is absent in the simple oval types.

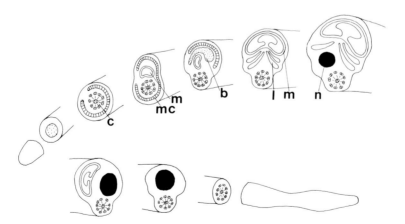

Fig. 2.11 Schematic diagram of structures of sperm, redrawn and much modified from B van Deurs (1972). b, beaded structure; c, centriolar structure; l, longitudinal sacs; m, mitochondrion; mc, membranous cap; n, nucleus.

The masses of sperm released when the seminal vesicles rupture may be held together by mucoid or gelatinous material (Reeve and Cosper 1975) but, since they are not enveloped by a membrane, most authors avoid the term spermatophore (Alvariño 1983*b*; Goto and Yoshida 1985; Nagasawa 1987*b*).

The fine structure of chaetognath spermatozoa (recently studied in *Spadella* by van Deurs 1972) is shown in Figure 2.11. The chaetognath spermatozoon is of the filiform type and its internal structure is unique to this phylum. Anteriorly there is a membranous cap (possibly an acrosome), some beaded sacs, and a centriolar structure. A very elongate mitochondrial derivative runs through most of the spermatozoon, and in its central portion there is a structure consisting of some longitudinal sacs lying next to the axonema. The nucleus is situated posteriorly, just anterior to a free flagellum (van Deurs 1972).

Chromosomes

In all the chaetognaths studied, the haploid number of chromosomes is 9 (Fig. 2.12).

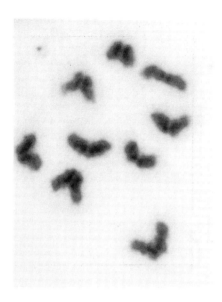

Fig. 2.12 Chromosomes of meiotic metaphase (*S. enflata*). Photograph by courtesy of C Thiriot-Quievreux and S Dallot.

Embryonic development

Both the planktonic eggs of *Sagitta* (without evident yolk material) and the attached eggs of *Spadella* (with uniformly distributed yolk granules) undergo the same total, equal, radial, indeterminate cleavage to form a blastula with a very small blastocoel. Gastrulation by invagination begins at around the 60-cell stage and ends in a classic gastrula (John 1933; Kuhl 1938; Hyman 1959; Kuhl and Kuhl 1965). The blastopore, which soon closes, marks the posterior end of the embryo. Anteriorly, two folds begin to develop and push into the archenteron. The primordial germ cells, having moved out of the endoderm, lie free in the archenteron in front of these two folds (Fig. 2.13(a)). They are marked by their size and by the germ cell determinant (Ghirardelli 1968), a special body of uncertain origin that can be traced back to the ova. Before the folds join the endoderm in the posterior region of the embryo, two coelomic sacs—the head coeloms—arise anteriorly and the invagination of the mouth and oesophagus appears. The primordial germ cells (two in *Sagitta*, one in *Spadella*) have divided once and half the resulting cells are left on either side of the folds (Fig. 2.13(b)). The definitive intestine and the

median mesentery in the tail arise from the inner walls of the folds; their outer walls, together with the inner archenteron walls, give rise to the mesoderm of trunk and tail. This formation of the gut anlage and embryonic coelom is unlike that found in any other phylum. As the embryo elongates, all cavities disappear and the anlagen of the brain and ventral ganglion are formed by proliferating ectodermal cells. Before hatching, the

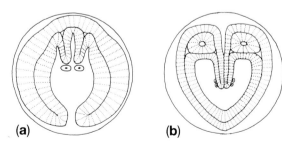

Fig. 2.13 (a) Formation of folds progressing into the archenteron, blastopore still open, (b) formation of head coelom and stomodaeum. Redrawn and modified from Burfield (1927).

embryo (approximately 1 mm long) is coiled within the egg so that the head overlaps the tail. Development time varies according to temperature but is between 19 and 68 hours (Dallot 1968; Reeve and Cosper 1975).

There is no metamorphosis after hatching, but rather a continued development of adult organs from embryonic structures. This takes about 6 days, so that, strictly speaking, it is incorrect to term the newly hatched chaetognath a larva, although many authors have done so (see Chapter 7). Thus, for example, at hatching the first beginnings of the lateral fins with some fin rays and a small tail fin are already present, extending and growing after hatching. Cavities appear again within the mesoderm and later in the intestine. Mesoderm cells begin to differentiate to form muscles in the head, trunk, and tail; lateral folds of the head ectoderm grow out to form the hood; all the other ectodermal structures are developed.

During the third or fourth day after hatching, the primordial germ cells, which are surrounded by meso-dermal cells separating the trunk and tail cavities migrate laterally from their position next to the intestine to the body walls, while the mesoderm cells differentiate to form the trunk–tail septum which separates those cells destined for the ovaries from those destined for the testes. At this stage, *Spadella* germ cells contain four nuclei. The development of the ovaries begins very late in ontogeny.

As the young chaetognath grows and elongates, its proportions come to resemble those of the adult more closely. There are differences between species in the growth and appearance of adult structures, for example, Reeve and Cosper (1975) reported hooks on the newly hatched *Sagitta hispida* and, according to Hyman (1959) and Kuhl (1938), hooks appear on the fifth day after hatching in *S. setosa*. However, such observations have been sporadic and embryology has been studied only in *Sagitta* and *Spadella*, so that much remains to be discovered in the embryonic and posthatching development of this phylum.

Acknowledgement

I am very grateful to H B Michel for her correction of the manuscript and to Q Bone for his help with the English formulation.

3

THE NERVOUS SYSTEM
Q Bone and T Goto

Chaetognaths are rapid, active predators feeding on living prey (Chapter 5) and it is not surprising that they have a relatively elaborate nervous system, informed by a variety of different sensory inputs, including near-field vibration receptors and (in most species) eyes of remarkably diverse structure.

In general, the central nervous system (Fig. 3.1) consists of six ganglia in the head and a large ventral ganglion in the trunk. A dorsal superficial cerebral ganglion in the head is linked by connectives to the ventral ganglion, and to paired vestibular ganglia on either side of the oesophagus. A short connective links the two vestibular ganglia with small oesophageal ganglia on either side of the oesophagus, and a caudal connective joins the rear of the vestibular ganglion as it loops around the back of the head. From the middle of this caudal connective a fine connective passes upwards to a very small suboesophageal ganglion, from where a ventral oesophageal nerve passes caudally along the gut. These arrangements have been most recently described by Goto and Yoshida (1987), who were the first to recognize the suboesophageal ganglion. In earlier texts, the connectives linking the ganglia of the head and the cerebral ganglia with the ventral ganglion were termed 'commissures', but as Goto and Yoshida (1987) have suggested, they are more properly termed 'connectives'. In addition to the connectives already mentioned, other nerves pass out of the ganglia (Fig. 3.1). These are either purely sensory, as, for example, the optic nerves and possibly the coronal nerves to the ciliary loop, or are assumed to be mainly motor, supplying, for example, the complex musculature of the teeth and grasping spines. It is not yet known whether proprioceptive endings exist in connection with the spine bases, or which nerves carry the axons of the various types of sensory cells on the head, so that although it seems probable that some of the cranial nerves may be purely motor, this remains to be demonstrated. In the case of the ventral ganglion, as will be seen below, all the nerves radiating from it are mixed nerves.

The basic plan of the ganglion with a central neuropile surrounded more or less completely by peripheral neuron somata, as described by Burfield (1927), differs for the small oesophageal ganglia and for the suboesophageal ganglion, where a neuropile region is absent, as Goto and Yoshida (1987) showed from serial semi-

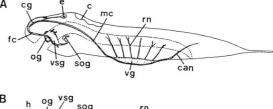

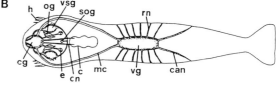

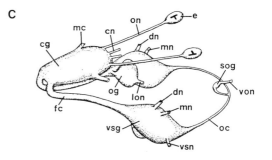

Fig. 3.1 General plan of chaetognath nervous system. (A) lateral view, (B) dorsal view, (C) anterodorsal view of head region. c, corona; can, caudal nerve; cg, cerebral ganglion; cn, coronal nerve; dn, dorsal nerve; e, eye; fc, frontal connective, h, hooks; lon, lateral oesophageal nerve; mc, main connective; mn, mandibular nerve; oc, oesophageal commissure; og, oesophageal ganglion; on, optic nerve; rn, radial nerve; sog, suboesophageal ganglion; vg, ventral ganglion; von, ventral oesophageal nerve; vsg, vestibular ganglion; vsn, vestibular nerve. From Goto and Yoshida (1987).

thin resin transverse sections. Most information about detailed structure is available for the cerebral and ventral ganglia but, as yet, very little is known of the connection of neurons within any of the ganglia, and of their functions. Some scattered observations give indications of the functional roles of the different parts of the chaetognath nervous system. Goto and Yoshida (1987) noted that after ablation of the cerebral ganglion, animals retain the ability to spread and close the grasping spines, which are thus, apparently, controlled by motoneurons within the vestibular ganglia. However, even for the ventral ganglion, which is the best known part of the nervous system, functional subdivisions remain to be established.

Cerebral ganglion

Histology

In the cerebral ganglion, as in the other ganglia, neuron somata are peripheral and do not form a complete cortex. Anteriorly, a neuropile region lies dorsally between two lateral masses of somata but for the greater part of the ganglion a central neuropile zone (Fig. 3.2) is enclosed dorsally and laterally by small somata (approximately 5 μm in diameter). A single pair of larger cells is found in the ganglion near the origin of the posterior connectives, their axons pass into the connectives. Immunocytochemical and supravital methylene blue studies have given some information about the course of the axons of different neuron somata, but unlike the ventral ganglion (see below), silver techniques have not yet been successfully applied to the cerebral ganglion. Nevertheless, it seems clear that, although they are present in the ventral ganglion, there are no large fibres within the cerebral ganglion. Bone *et al.* (1987*b*) have shown that much of the central neuropile is positive to antisera raised against the molluscan neuropeptide RFNH2 (Greenberg and Price 1980) and have identified certain positive neuron

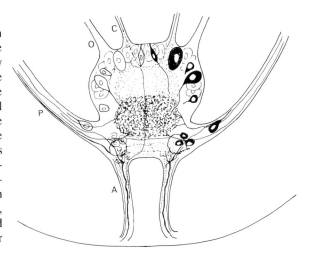

Fig. 3.3 Schematic dorsal view of cerebral ganglion (anterior below) showing central neuropile and dark cells (on right of diagram) positive for RFNH2-like material. The two large cells at the entrance of the anterior connectives are not RFNH2-positive. a, anterior connective; c, coronal nerve; o, optic nerve; p, posterior connective. From Bone *et al.* (1987).

Fig. 3.2 Transverse section of cerebral ganglion in *Spadella schizoptera* showing central neuropile (N) and lateral and dorsal cell bodies. Semithin resin section, toluidine blue. Scale bar: 20 μm.

somata in constant positions within the ganglion (Fig. 3.3). Axons from some of these cells pass caudally to the ventral ganglion and others pass anteroventrally to the vestibular ganglion. There are no synapses on the neuron somata, all synaptic contacts take place in the neuropile but, unfortunately, nothing is known of connections within it. Goto and Yoshida (1987) have observed 5-hydroxytryptamine-like positive cells, and a plexus (less dense than the RFNH2-like plexus) of similar positive fibres within the central neuropile. However, unlike the RFNH2-like fibres, the 5-HT-like positive fibres are confined to the ganglion and are not found in the connectives or in nerves linked with the ganglion.

The enigmatic paired retrocerebral organs (see

below) lie dorsolaterally in the mid-region of the ganglion.

Ultrastructure

Ultrastructural observations by Rehkämper and Welsch (1985) have shown that the cerebral ganglion is invested in a multilayered sheath (Fig. 3.4), which is,

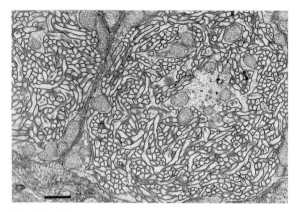

Fig. 3.5 Cells of the retrocerebral organs in *S. crassa* showing intertwining microvilli. Scale bar: 1 μm.

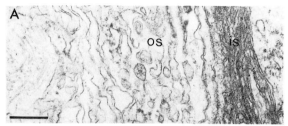

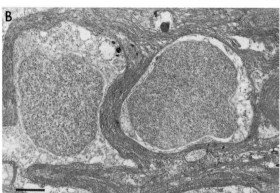

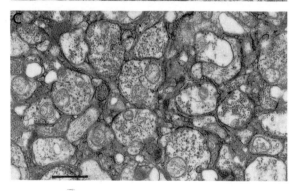

Fig. 3.4 (A) The inner (is) and outer (os) sheath of the cerebral ganglion in *S. setosa*. Scale bar: 0.5 μm. (From Rehkämper and Welsch (1985) with permission). (B) Inner myelin-like layer of sheath investing cerebral ganglion somata in *S. hexaptera*. Scale bar: 2 μm. (C) Neuropile region of cerebral ganglion showing profiles containing electron-lucent and dense-cored vesicles. Scale bar; 1 μm.

in some respects, reminiscent of the vertebrate myelin sheath. The outer zone of the sheath consists of flattened cells with few organelles, occasionally connected by what appear to be gap junctions and the inner layer is made up of densely-packed thin lamellar processes separated by electron-dense material (Fig. 3.4(A)). This inner part of the sheath extends into the ganglion to invest the neuron somata, as shown particularly clearly in *Sagitta hexaptera* (Fig. 3.4(B)) by Goto and Yoshida (1987). The resemblance to myelin is evident. However, this presumably glial sheath does not extend into the neuropile region of the ganglion, where many synaptic contacts, containing both dense-cored and electron-lucent vesicles, are found (Fig. 3.4(C)). No electrical synapses (gap junctions) have been found in the cerebral ganglion or elsewhere in the nervous system. As in the ventral ganglion, most of the fibre profiles in the neuropile of the cerebral ganglion are closely apposed without any intervening glial material.

The paired retrocerebral organs, which lie laterally at the caudal end of the ganglion and open to the exterior via fine canals terminating in a single pore, were first observed by Kowalewsky (1871), and subsequently examined by Grassi (1883), who supposed that they might be sensory organs. Scharrer (1965) and Goto and Yoshida (1987) examined their ultrastructure (Fig. 3.5) and showed that they consisted of large cells with evaginated outer membranes forming an intertwining mass of long microvilli, each with a central dense thread. A similar organ is found in the equally transparent crustacean *Leptodora*, but in neither group is there any indication of its function.

Physiology

Goto and Yoshida (1987) have shown spontaneous activity (spikes of different amplitudes recorded extracellularly by KCl-filled microelectrodes) in the cerebral ganglion of *Spadella schizoptera*, but changes in such activity resulting from sensory input or motor activity have not been investigated. From the immunocytochemical studies, it seems clear that 5-HT-like positive neurons are associative or coordinating, since no positive fibres have been found in the nerves connected to the ganglion. On the other hand, RFNH2-like positive fibres pass out of and enter the ganglion in the posterior connectives linked to the ventral ganglion, and a conspicuous pair of positive fibres enters the ganglion from the vestibular ganglion. A single pair of positive fibres passes out in the coronal nerves but does not innervate the corona because it continues caudally beyond the corona. From these data it seems that some of the RFNH2-like positive neurons of the ganglion are also likely to be coordinating rather than motor.

Ventral ganglion

Histology

Both Hertwig (1880*a,b*) and Grassi (1883) recognized that the elongate ventral ganglion was built upon an essentially scalariform plan, but it is only recently that details of the arrangement have been demonstrated by the use of specific methods for nerves, and by immunocytochemical staining for neuropeptides and putative neurotransmitters (Bone and Pulsford 1984; Bone *et al.* 1987*b*; Goto and Yoshida 1987; Goto unpublished). Although as yet there is only a partial correlation between the architecture revealed by methylene blue and reduced silver techniques and that revealed by antisera to different neuropeptides, it is evident that the ventral ganglion has a strikingly serial arrangement which is constant within the species (and probably varies little between species). The ganglion is a long rectangle lying on the ventral surface of the trunk about one-third of the body length from the head; it is covered by a thin layer of epidermal cells. A single species, *Bathybelos typhlops*, has been reported as lacking a ventral ganglion in the usual position and has instead a so-called 'cerebral' ganglion on the dorsal side of the neck between head and trunk. This resembles the normal ventral ganglion in shape and size (Owre 1973). The corona ciliata which occupies this position in other species is absent in *Bathybelos*. It is very curious that such a stereotyped feature as the ventral ganglion should be lacking or so altered in position; perhaps it is related to the notable reduction of other features such as the longitudinal musculature and fin rays, suggesting that the species is relatively inactive. The special features of *Bathybelos* have been re-examined by Bieri (personal communication) who concludes that they are not the result of misinterpretation of a damaged specimen.

In all other species, the ganglion consists of a central fibre zone flanked by neuron somata. Occasional cell bodies lie dorsally in the mid-line; these have been interpreted as glial cells (Bone and Pulsford 1984), but further caudally there seem to be neuron somata in this position (Goto and Yoshida 1987). Twelve pairs of nerves radiate laterally along the length of the ganglion and two large connectives link the anterior end of the ganglion with the cerebral ganglion. Two main nerve trunks pass out of the hind end of the ganglion towards the tail; Figures 3.6 and 3.7 show the general arrange-

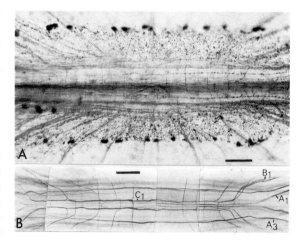

Fig. 3.6 (A) Entire ventral ganglion of *S. setosa* (anterior to right) stained supravitally with methylene blue showing lateral cell bodies and central neuropile zone traversed by axons crossing to the opposite side of the neuropile. Scale bar: 50 μm. (B) Similar view from fixed ganglion impregnated with silver to show larger fibres in central neuropile tracts. Fibres labelled correspond to those shown in **Fig. 3.7**. Scale bar: 50 μm.

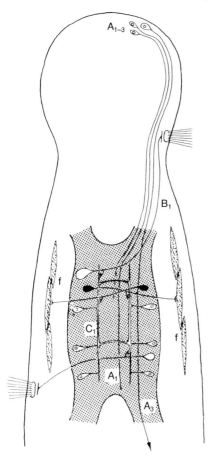

Fig. 3.7 Schematic diagram of larger fibres of ventral ganglion in *S. setosa* seen in Fig. 3.6(B). Cell bodies are drawn according to their functions (white, sensory; black, motor; nucleated, interneurons). (From Bone and Pulsford 1984.)

ment. Work in the past few years has shown that within the central neuropile the ventral ganglion contains several longitudinal fibre tracts derived from fibres passing to the ventral ganglion from the cerebral ganglion and from the laterally-placed cell bodies within the ganglion itself. Some of the fibres in the longitudinal tracts are likely to be associative or coordinating, others are thought to be sensory, passing to the ciliary vibration receptors of the trunk and rear of the head, and others are presumed to be motor to the trunk musculature. The nerves radiating from the ganglion along its length are mixed nerves, containing both sensory and motor fibres and, although details of the way in which they are connected with the longitudinal fibre tracts remain to be worked out, it seems clear that the majority of centrifugal fibres in the mixed nerves from the ganglion are derived from axons that run for a varying distance in one of the longitudinal tracts and then cross to exit via nerves contralateral to the cell body. Synaptic connections between the longitudinal fibre tracts and crossing fibres exist (see below) but nothing is definitely known of their relationship with neuron somata in the ganglion. The rounded neuron somata themselves (which are occasionally up to 10 μm in diameter, but average 5 μm) do not receive synapses, all connections being made within the central neuropile into which their axons pass and through which the longitudinal tracts run. Among these tracts are a small number of large fibres (some up to 6 μm in diameter in silver-stained preparations) which are constant in number and position in different specimens, and probably between different species. In *Sagitta enflata*, for example, the arrangement of the larger fibres is strikingly similar to that in *Sagitta setosa*. Some of these can be recognized both in silver preparations and in those visualizing RFNH2-like material; a single pair are visualized with antisera to 5-HT (Fig.3.8). Neurotensin-like and (rarely) metenkephalin-like positive fibres and somata are present in the ventral ganglion (Goto and Yoshida 1987) but neither material is found in any of the larger fibres. The large fibres are paired and linked by transverse connections and most terminate within the ventral ganglion, whether they are formed by cells within it or are derived from somata in the cerebral ganglion. However, one pair of large fibres, A3, enters the ganglion anteriorly from the cerebral ganglion and runs throughout its length to pass out via the posterior nerve trunks.

The A series of large fibres (A1–3) are evidently single axons, but it is not yet clear whether this is also the case for the C1 pair of fibres. These C1 fibres are the most conspicuous fibres in the ganglion. They lie entirely within it and taper at both ends. They are connected by transverse bridges at constant positions along their length and give off side branches transversely. Some of these lateral branches (which have not been traced beyond the borders of the central neuropile zone) may be the axons of somata within the ganglion that contribute to the C1 pair. The C1 pair may be 'giant' fibres formed by the fusion of several axons or, although apparently a single fibre in silver preparations, may actually consist of a number of smaller separate fibres closely applied to one another. At either end of the ganglion there is a pair of larger neuron somata whose axons pass out of the anterior and posterior main nerve trunks after crossing the

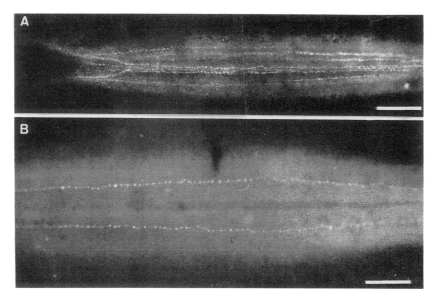

Fig. 3.8 (A) Ventral ganglion of *S. setosa* stained for RFNH2-like material. Scale: 100 μm. (B) Single pair of 5-HT-like positive fibres in *S. cephaloptera*. Scale bar: 100 μm.

central neuropile zone, as seen in Figure 3.9. These pass directly out in the connectives and travel anteriorly to the cerebral ganglion. Those at the caudal end of the ganglion loop across to run longitudinally for a short distance before passing out of the ganglion. The course of the posterior axons has not been traced, but

the anterior axons run in the connective until they reach the posterior edge of the head (giving off a small side branch on their course) and terminate at the base of the ciliary fan receptor behind the head. These axons contain RFNH2-like material (Fig. 3.8), and are the only axons in the system which have been traced to their termination at the periphery from the cells of origin. It is evident that further progress in tracing neuron connections and the course of their axons within the ganglion will require such techniques as Co^{2+} back-filling and improved immunocytochemical studies.

Ultrastructure

Although ultrastructural observations (Bone and Pulsford 1984; Goto and Yoshida 1987) have shown some interesting features of the organization of the ganglion, it has proved difficult to correlate these with what is known at the light microscope level. The chief features of the system at the ultrastructural level are the close apposition of the fibres in the neuropile (most of which are 1 μm or less in diameter) without intervening glial processes and the abundance of synaptic contacts. All of the synaptic contacts so far observed contain electron-lucent vesicles some 40 nm in diameter. Some infrequent fibre profiles contain larger (80–90 nm dense-cored vesicles but synapses with vesicles of this type have not been seen. Synaptic contacts occur between longitudinally-running fibres; between longi-

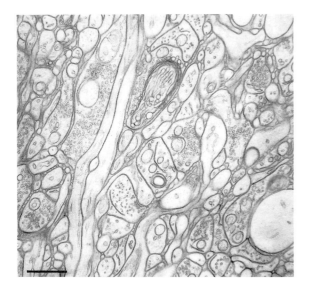

Fig. 3.9 Transverse section of ventral ganglion neuropile showing close apposition of fibres and absence of obvious glial elements between many fibre profiles. Scale bar: 0.5 μm.

tudinally-running and transversely running fibres
(where the longitudinal fibres are presynaptic to the
transverse fibres); and between transverse fibres and
smaller longitudinal fibres (where the transverse fibres
are presynaptic). Some glial profiles are seen in sections
of the ganglion and there are a number of cell bodies
superficially above the central neuropile zone which
have been interpreted as glial cells, but have also been
regarded as neuron somata. Nevertheless, the main
impression gained from examination of such fields is
that the majority of fibres are apposed without being
separated by glial material. In this respect the ventral
ganglion differs from the cerebral ganglion, where
Rehkämper and Welsch (1985) describe numerous
profiles with dense granules that possibly belong to
specialized glial cells.

Fig. 3.10 Extracellular suction electrode record of rhythmic
activity from isolated ventral ganglion of *S. elegans*. Time
marker, s. (From Bone and Pulsford 1984.)

Physiology

Physiological studies on the ventral ganglion have been
confined to extracellular suction electrode records of
activity (Fig. 3.10) because although the ganglion is
easily accessible as a consequence of its superficial
position, cell bodies, and the fibres within the neuropile
are small. The rhythmic activity recorded from the
isolated ganglion is similar in frequency to that of the
intact animal, and radial nerves in a pinned-out
preparation stimulated via a small suction electrode
evoke contractions of regions of the trunk musculature.
It seems clear, therefore, that the trunk musculature is
innervated by neuron somata within the ventral ganglion and that the regular rhythmic activity of the
animals examined (*Sagitta setosa* and *Sagitta elegans*,
see Chapter 4) is controlled by a system intrinsic to the
ganglion. At present, it is only possible to suggest that
some of the larger fibres within the ganglion have a
coordinating function, but any functional interpretation
of the organization of the ganglion (seen in part in Figs
3.6 and 3.7) is premature.

Epidermal nerve plexus

There is an extensive epidermal nerve plexus in the
trunk consisting of a network of fine nerve fibres lying
embedded within the bases of the epidermal cells
adjacent to the basement membrane. These fibres are
derived from the radial nerves of the ventral ganglion.
Many (but probably by no means all) of these fibres
contain RFNH2-like material; others are presumably
motor to the trunk musculature (and hence probably
cholinergic (see Chapter 4). Other fibres within the
plexus react with antisera to glutamate, galanine, beta-
endorphin, and aspartate (Duvert, personal communication). It is not known whether the fibres positive
for RFNH2-like material are motor or sensory. Hertwig (1880*a,c*) observed large bipolar ganglion cells
amongst the fibres of the epidermal plexus; these have
more recently been observed in silver and methylene
blue preparations—they do not contain RFNH2-like
material. Nothing is known of the function of the fibres
within the epidermal plexus that are not motor to the
trunk musculature but it seems obvious that some of
them must innervate the ciliary fence receptors (see
below), although they are so numerous that not all of
them can do so. Presumably they are touch-sensitive.

Visceral nerves

A ventral visceral nerve arises from the suboesophageal
ganglion—a small group of neuron somata under the
oesophagus (Goto and Yoshida 1987) (see Fig. 3.1).
After its origin from the ganglion the nerve divides to
form the dorsal and ventral visceral nerves, passing
along the gut in the mesenteries linking the gut with
the body wall. In each nerve (in *Sagitta setosa* and
Sagitta elegans) there are some 14–22 fibres, varying in

diameter and sometimes containing electron-lucent vesicles. It is not known if the fibres branch as they pass along the gut. Presumably they innervate the smooth muscle cells of the gut, but no nerve terminals have yet been observed, nor is it known if the nerves continue beyond the anus (e.g. innervate the gonads).

Receptors

The most conspicuous receptors in almost all chaetognaths are the paired eyes (differing remarkably in structure between species; lacking in a few) and the ciliary fence receptors arrayed on the head trunk and tail. The function of these are fairly clear from their morphology, but there are a number of other assumed receptors whose functions are not evident.

Eyes

Most adult chaetognaths possess paired eyes on the dorsal surface of the head, although in some species they are lacking (for example, all *Heterokrohnia* species and some *Eukrohnia* species). Each eye is rounded, dorsoventrally flattened, and enclosed in a capsule penetrated anteriorly by the optic nerve. In most species the eyes are easily seen as a pair of pigmented spots whose size and shape differs, so that they may be useful in species identification. Eye morphology has mainly been studied in epipelagic *Sagitta* species with light microscopy (Burfield 1927) and, more recently, by electron microscopy (Eakin and Westfall 1964; Ducret 1975, 1978; Goto and Yoshida 1983; Goto *et al*. 1984, 1989). Figure 3.11 shows the general appearance of a transverse section near the mid-line of the eye in *Sagitta hexaptera*. The pigment spot is a single pigment cell in the centre of the eye. It has a rather complicated topology which gives a T-shaped profile (Fig. 3.12) when cut transversely just anterior to the mid-region of the eye. The shape and size of the pigment granules are highly variable. The photoreceptor cells are arrayed around the pigment cell and are divided into three parts: the receptor process, the cell body, and the axon (Fig. 3.11(b)). The number of photoreceptors differs with the species and probably depends on the stage of growth; in *Sagitta pacifica* there are fewer than 100 and in *S. hexaptera* there are more than 400. The receptor process is ciliary and consists of the distal segment and conical body. Goto and Yoshida (1988) have demonstrated (using a histochemical method) rhodopsin-like material in the distal segment, suggesting that this is the primary photoreceptive site. The conical body is refractive in living preparations and hence has been supposed to be a dioptric system of some kind. It is rich in glycogen, as is the cone paraboloid of vertebrate cones which it superficially resembles (Goto and Yoshida 1984), but its true function is unknown. In epipelagic *Sagitta* species the distal process (the presumed photoreceptive region) is next to the pigment cell, and the cell body is located near the periphery of the eye, therefore the eye is inverted. Eakin and Westfall (1964) described the distal segment as composed of an array of long, very narrow tubules enclosed in an extension of the ciliary membrane but later work by Goto *et al*. (1984) showed that it is, in fact, not tubular but composed of a stack of lamellae perforated by numerous pores. Such a perforated lamellar structure is unique amongst photoreceptors.

Goto *et al*. (1989) examined ten species of *Sagitta* from three different habitats, finding variations in the shape of the pigment cell and in the arrangement of the photoreceptive region (PR) in the different species, related to their habitat. They distinguished five eye types (Fig. 3.12). In types I and II there is a single large pigment cell surrounded by a wide area of the PR, termed the central PR. In addition, the type II eye has masses of PR near the periphery, termed the peripheral PR. Type III eyes have small pigment cells and little central PR, with widely distributed peripheral PR. In the type IV eye, the pigment cell is also small and the central PR extends to the periphery. The pigment cell in the type V eye is absent and the PR occupies a wide area of the dorsal half of the eye. Type I and II eyes were found mainly in epipelagic species, types III and IV in mesopelagic species, and type V in bathypelagic species. Presumably, these difference are related to light levels in the different habitats. Those species which occur in the mesopelagic and bathypelagic zones are not degenerate but appear best adapted for photon capture.

Eye structure has also been examined in *Spadella* (Goto and Yoshida 1988) and *Eukrohnia* (Ducret 1975, 1978; Goto, unpublished data). The eyes of *Spadella schizoptera* (the Japanese specimens examined have recently been separated as *Paraspadella gotoi*, Casanova 1990) are similar to those of epipelagic *Sagitta*,

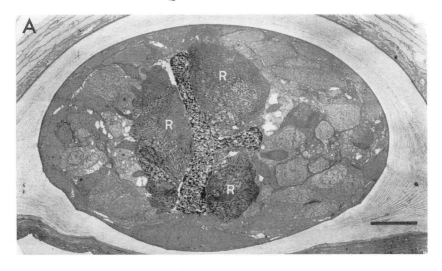

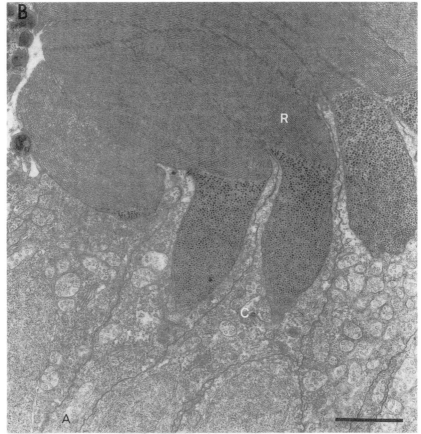

Fig. 3.11 (A) Transverse section near mid-region of eye in *S. hexaptera*. A central pigment cell (P) separates the three receptor process regions (R) of the receptor cells. Scale bar: 10 μm. (B) Photoreceptor cells from *S. hexaptera*. R, receptor process; C, cell body; A, axon. Scale bar: 2 μm.

Eye type

Species

epipelagic

mesopelagic
bathypelagic*

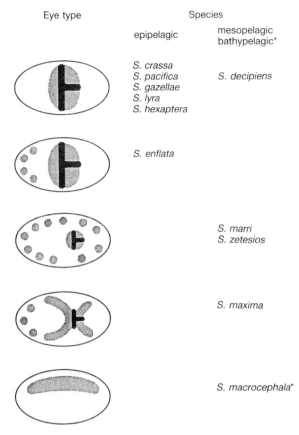

S. crassa
S. pacifica   S. decipiens
S. gazellae
S. lyra
S. hexaptera

S. enflata

S. marri
S. zetesios

S. maxima

S. macrocephala*

Fig. 3.12 Different types of eyes in *Sagitta* species from different habitats. Pigment cells, black; photoreceptor regions of receptor cells, hatched. (Goto *et al.* 1989.)

except that the pigment cell granules are rod-shaped. However, not only are the eyes of different *Eukrohnia* species different, but all are entirely different to those of *Sagitta* and *Spadella*. The morphology is well known in *E. hamata*, (Fig. 3.13) where there are around 100 lens-like structures seen in dorsal view; these are usually termed 'ommatidia'. In transverse section the presumed photoreceptor cells are arranged with the 'ommatidia' pointing outwards (Fig. 3.13), each located at the distal end of a single photoreceptor cell. The 'ommatidium' is thus clearly different from those of the arthropod compound eye and the photoreceptor cell consists of the 'ommatidium', connecting cilium, cell body, and axon. The presumed photoreceptor region is located around the 'ommatidium'. Ducret (1978) regarded the photoreceptive region as composed of simple microvilli, and called it a 'rhabdome', but it is in fact very complicated and differs from the perforated lamellae seen in *Sagitta*. In *E. fowleri*, the pigment cell enlarges at night (Ducret 1975) but it has yet to be examined in detail. *E. kitoui* and *E. proboscidea* have a pigmented region but lack 'ommatidia'; again, their eyes have not been examined in detail.

From a survey of eye structure in many species, Ducret (1978) proposed that from an initial inverted type of eye, the receptor cells changed to an orthotopic orientation in which the 'ommatidia' were developed (Fig 3.14 and Table 3.1).

Although there has been no electrophysiological work on photoreception, some interesting behavioural studies have been carried out. Positive phototaxis has been observed in *Sagitta* by Esterly (1919); Pearre

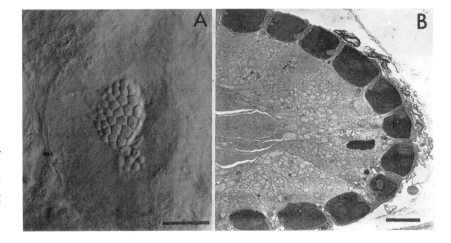

Fig. 3.13 (A) Lens-like 'ommatidia' of the eye of *Eukrohnia hamata*. Scale bar: 25 μm. (B) Eye of *E. hamata* in transverse section, showing 'ommatidia' (dark) at the outer ends of each receptor cell. Scale bar: 5 μm.

Table 3.1 Eye structure in chaetognaths

Character	Sagitta	Eukrohnia	Heterokrohnia	Krohnitta	Pterosagitta	Spadella	Bathyspadella	Krohnittella	Bathybelos
Pigment cell									
present	most species	*fowleri* *proboscidea* *kitoui*		*subtilis* *pacifica*	*draco*	most species			
absent	*macrocephala*	*bathyantarctica*					*edentata*	*boureei* *tokiokai*	
Ommatidia		*hamata* *bathypelagica* *minuta*							
Photoreceptive cells									
inverted	most species				*draco*	*cephaloptera* *schizoptera*			
semi-inverted		*fowleri* *proboscidea* *hamata* *bathypelagica*							
orthotopic	*macrocephala*	*macroneura* *flaccicoeca*							
No eye			all species			*anops*			*typhlops*

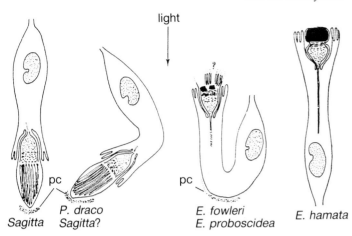

light

pc

Sagitta

P. draco
Sagitta?

pc

E. fowleri
E. proboscidea

E. hamata

Fig. 3.14 Diagram showing possible transformation of inverted photoreceptors on left to orthotopic receptors with 'ommatidia' on right. pc, pigment cell. After Ducret (1978).

(1973); Goto and Yoshida (1981, 1983) and Sweatt and Forward (1985*a*,*b*). During phototaxis, the swimming pattern is the characteristic dart and sink motion (see Chapter 4). Using phototaxis as a behavioural measure of photosensitivity, Sweatt and Forward (1985*b*) found threshold light intensity in *Sagitta hispida* to be approximately 1013 photons m^2 s^{-1}, and that spectral sensitivity was maximal at 500 nm, with a small shoulder at 600–620 nm. From this they inferred the presence of both a rhodopsin-like pigment and of another pigment with an absorbance maximum around 600 nm. Goto and Yoshida (1981, 1983) saw another photic response in *Sagitta crassa*, this was also observed in *Sagitta hispida* by Sweatt and Forward (1985*a*,*b*). The animals swam rapidly towards the light source in response to mechanical shock or to sudden reduction in light intensity. This kind of light-adapted startle response is called quick target-aiming behaviour. By unilateral blinding and using two-light stimuli, Goto and Yoshida (1983) concluded that both reactions were telephototaxes. The morphological basis for the target locating ability may lie in the shielding effects of the complex pigment cell shape.

Ciliary fence receptors

These consist of fans of elongate cilia on the head, trunk and tail. In *Spadella cephaloptera* (Fig. 3.15) the cilia are up to 100 μm long. In other species the cilia may be much longer, reaching a maximum in *Pterosagitta draco*, where the paired long fans on the anterior part of the trunk are some 200 μm long. Feigenbaum (1978) has examined the pattern of the ciliary fence receptors in several species and has shown that there is greater variability in their array in *S. elegans* than in *S. hispida* (where they are so conspicuous as to merit the specific name) or *S. enflata*. In the last species, Feigenbaum observed about 600 organs on an 18 mm adult. Each cilium takes origin from one of a set of closely packed cells and is surrounded by a corolla of microvilli. Long ciliary rootlets pass into the basal portion of the ciliated cells but these are non-motile and evidently fairly stiff in life, although in SEM views they are somewhat tangled during preparation (Fig. 3.16). Reisinger (1969) and Welsch and Storch (1983*b*) have shown that these are secondary sensory cells receiving synapses (containing electron-lucent vesicles) at their bases from central axons (a condition atypical for invertebrates, although known also in mechanoreceptors of larvacean tunicates and cephalopods). This conclusion is supported by the innervation of the pair of ciliary fence organs at the back of the head where RFNH2-like positive fibres can be seen to pass from neuron somata in the ventral ganglion to the base of the receptor fans. On the trunk and tail the sensory cells are set either at right angles to the long axis of the body or parallel to it; the majority being at right angles. In *Spadella cephaloptera*, for example, there are approximately 32 receptor organs on each side of the

Fig. 3.15 Ventral view of *Spadella cephaloptera* showing ciliary fence organs (white spots) on head, trunk and tail fin— SEM. Scale bar: 400 μm. (From Bone and Pulsford 1978.)

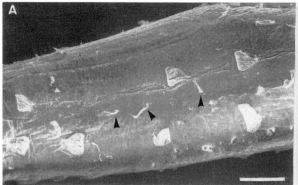

Fig. 3.16 (A) Longitudinal and transverse orientation of ciliary fences in *S. elegans*. Note much wider fences transverse to long axis of body than those parallel to axis (arrowed)—SEM. Scale bar: 100 μm. (B) Side view of ciliary fence—SEM. Scale bar: 10 μm.

body, of which eight are aligned parallel to the long axis. Since the length of the fences indicates the number of cells in each receptor, and since those fences transverse to the long axis are normally longer (Fig. 3.16(A)) there are more receptor elements aligned across the long axis than the number of organs indicates. Although direct proof by means of records of activity from the receptors themselves is not available, it seems certain that the ciliary fence receptors are involved in the response to short range vibrations. Horridge and Boulton (1967) and Feigenbaum and Reeve (1977) have shown that both *Spadella* and *Sagitta* will attack small vibrating probes placed at distances up to 3 mm from the animal. Feigenbaum and Reeve found that *Spadella cephaloptera* and *Sagitta hispida* responded at all frequencies tested, but the response curves (Fig. 3.17) differed; that for *S. schiz-*

optera peaking at 30 Hz and that for *S. hispida* peaking at 150 Hz. Earlier, Horridge and Boulton (1967) had found that the response curve for *S. cephaloptera* peaked at 12 Hz. The lower frequency responses appear to be within the range of vibrations produced by swimming copepods, but 150 Hz is higher than any reported from copepods, as is the peak of 60 Hz observed in the response curve of *Sagitta euneritica* by Cummings (in Feigenbaum and Reeve 1977). Presumably the ciliary fence receptors respond chiefly to deformation by water movements in the plane transverse to the axis of the organ, and their orientation on the body reflects the fact that vibrating prey will be more accessible ahead of the animal than to the side.

Other ciliated receptors

Five other types of ciliated cells which are likely to be receptors have been described. In *S. cephaloptera*, Ahnelt (1984) briefly described smaller groups of ciliated cells which had shorter, stiffer cilia than the ciliary fences. In the same species Bone and Pulsford (1978) observed single large, apparently stiff, cilia projecting from the surface of the head and caudal region. These are about twice the diameter of the cilia in the fence organs and contain large numbers of tubules. Similar large cilia with multiple tubules are found in larvaceans, where they are known experimentally to be touch receptors (Bone and Ryan 1979), and this is their presumed function in *Spadella*. Also in *Spadella* on the ventral surface of the head, the same authors observed curious large, domed cells covered by cuticle, with an invaginated tube in which lay two cilia. The cilia did not emerge from the tube as far as could be observed. Perhaps such enclosed ciliary slit receptors are chemoreceptors? Neither of these two

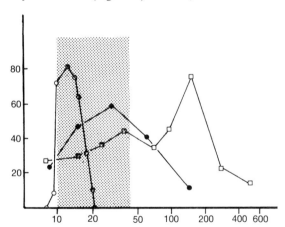

Fig. 3.17 Response curves for attack on vibrating probes by *Spadella cephaloptera* (open circles), *S. schizoptera* (filled circles), and *Sagitta hispida* (squares). The stippled zone represents the range of vibrations produced by copepods (Newbury 1972). From Feigenbaum and Reeve (1977).

presumed receptor types has been observed in other chaetognaths.

In *Sagitta setosa* and other *Sagitta* species, the cilia of presumed receptor cells protrude from the apertures at the tips of the papillae along the vestibular ridge on the ventral surface of the head (Bone and Pulsford 1984; Ahnelt 1984). Like the single large cilia of the head and caudal region in *Spadella*, these cilia have multiple tubules. In methylene blue supravitally-stained preparations, spindle-shaped cells with long basal processes are seen in the vestibular papillae, and it seems likely that, unlike the ciliary fence receptors, these are primary sensory cells. Other similar cells which are seen in methylene blue preparations on the ventral surface of the head laterally to the mouth are probably ciliated receptor cells, whose cilia emerge via small pits in the ventral cuticular plate.

Lastly, although there have long been suggestions that the ciliary loop may be a receptor, and that a coronal nerve passing to the cerebral ganglion is linked

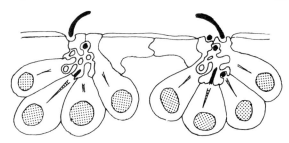

Fig. 3.18 Schematic transverse section of corona ciliata in *S. glacialis* showing putative ciliated sensory cells. After Malakhov and Frid (1982).

with it, the first description at the ultrastructural level of probable receptor cells in the organ has been given by Malakhov and Frid (1984). These authors describe primary receptor cells with branching cilia (Fig. 3.18). Unfortunately their account is brief and further investigations of this enigmatic organ are required.

Conclusion

Chaetognaths have a complex nervous system which is largely epidermal. They have unusual eyes and vibration receptors, as well as a suite of other putative receptors whose function is unclear. Future tracer and immunocytochemical studies, combined with behavioural experiments, are likely to reveal something of the functional subdivisions of the nervous system and to clarify the role of the different sense organs, but neither these, nor the application of modern electrophysiological methods to the unsuitably small neurons of the group, seem likely to throw any light on its affinities. The organization of the nervous system is (tantalizingly) in several respects both different and similar to that of other groups, as Grassi (1883) recognized. However, since it still remains poorly known, it is premature to use the system in speculation about the relationships with other groups. As in the other characteristics of the chaetognaths, the morphology and histology of the nervous system (insofar as it is known) seems very stereotyped and similar between different species, and hence is of little value in determining relationships within the group. There is, however, a striking exception to this generalization, in that the structure of the eyes is so diverse that it may be useful in distinguishing species and, perhaps, in recognizing generic affinities.

4

LOCOMOTION AND BUOYANCY
Q Bone and M Duvert

Chaetognaths can produce very rapid, directed movements to catch their prey and to avoid predators, and many (perhaps most) of the pelagic species that are denser than sea water show regular brief bursts of swimming alternated with passive sinking. This rhythmic hop-and-sink pattern of activity enables them to search through the water for their prey (see Chapter 5). Other species, such as *Sagitta enflata*, are either neutrally buoyant or close to neutral buoyancy, and seem less active, although *S. elegans* (which is close to neutral buoyancy) shows the same kind of rhythmic activity pattern as a dense species like *S. setosa* (Bone *et al.* 1987a). The locomotor muscles of the trunk which oscillate the body in the vertical plane during swimming have an unusual and striking structure and include several different fibre types, one type being unique to chaetognaths.

Hydroskeleton

The chaetognath trunk contains no organs apart from the intestine and the gonads (Chapter 2); the epithelial layers of the body are built around a hydroskeleton that plays a fundamental role in the organization of the trunk and tail. The two main components of the hydroskeleton are the basal membrane and the fluid contained within the body cavity (Chapter 2).

As in all hydroskeletons, internal pressure must be controlled for the optimal operation of the trunk musculature; pressures up to 1 kPa have been recorded from the body cavity during locomotion (Bone *et al.* 1987b).

Muscle structure

Primary musculature

Muscle fibre structure and arrangement

The primary musculature, containing two muscle fibre types (A and B), is the basic tissue of the body wall and makes up approximately 80 per cent of the trunk wall volume. Since chaetognaths appear to grow throughout life, regardless of age, there is a strict correlation between the size of the hydroskeleton and the volume of the primary musculature. As Prenant (1912) first suggested, during this continued growth new muscle fibres are added to those already existing (Dress and Duvert 1983; Duvert, Dress and Salat, 1987), and each of the four quadrants into which the primary musculature is divided grow by the regular production of groups of fibres. Nothing is known of the mechanisms controlling this parsimonious and regular growth. Each quadrant contains one myogenic site, lying at its edge, next to the lateral fields. These produce, alternately and asynchronously, two cell types, giving rise to the two types of muscle fibre which Burfield (1927) observed. The first myoblast type gives rise to a fixed number of cells which remain grouped, leave the cell cycle and differentiate into A fibres. The second myoblast type differentiates similarly, but later, and gives rise to B fibres.

This process of muscle fibre formation is extremely regular, occurring in the trunk in such a way that three groups of A fibres are produced as two groups of B fibres are differentiated (Duvert and Dress, unpublished data). In the tail, single groups of A fibres are separated by two groups of B fibres. These groups of A and B fibres remain ordered, so that their relative disposition not only determines the architecture of the muscle but also reflects its ontogenesis. At the same time, the regular grouping of A and B fibres (Fig. 4.1(A) and (B)) suggests that these groups of fibres could perhaps be considered as mechanical units, each possibly being some kind of 'motor unit'. Curiously, B fibres are apparently lacking from the locomotor musculature of *Spadella* (Duvert, unpublished data).

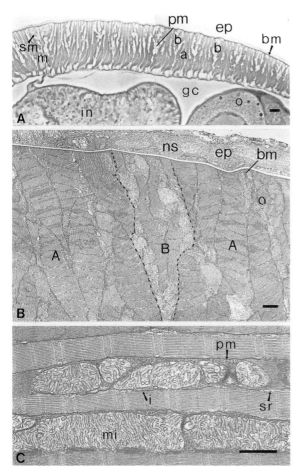

Fig. 4.1 (A) Transverse section of part of trunk wall of *S. setosa* showing regular alternation of A and B fibres in the primary muscle (pm), and the position of the secondary muscle (sm) near the mesentery (m). bm, basement membrane; c, body cavity; ep: epidermis; in, intestine; o, ovary. Semithin resin section, toluidine blue. Scale bar: 10 μm. (B) Survey electron micrograph of transverse section of the trunk musculature near the epidermis at the basement membrane level. A group of B fibres is outlined. bm: basement membrane; ep: epidermis; ns: nerve bundles. Scale bar: 1 μm. (C) Longitudinal section of a B fibre showing mitochondrial columns (mi) (cryopreservation). i, invaginations; pm, plasma membrane; sr, sarcoplasmic reticulum. Scale bar: 1 μm.

Both types of fibre have the same contractile apparatus and both exhibit a pronounced myofibrillar ATPase activity over a very wide pH range (3.5–9.0); ATPase activity does not, therefore, differentiate the two fibre types, nor indicate their ontogenetic stages. Nevertheless, certain features distinguish the two:

Form and position The B fibres are arranged longitudinally along the basement membrane and each fibre is inserted into the membrane at regular intervals. The A fibres make up longitudinal oblique arrays. These are parallel and are arranged helicoidally, alternately left and right handed. It is this arrangement which gives the zigzag appearance of the living muscle sheet when the animal is opened and pinned out for experiments. The A fibres insert on the basement membrane by a tendinous end and the B fibres form groups that are intercalated between two neighbouring groups of A fibres (Fig. 4.1(B)). The nuclear regions of the latter extend to the border of the body cavity; this never occurs in more superficial B fibres. This complex architecture must presumably be related to the kinds of movements made by the animal, but these are not known in detail.

Ultrastructure The A fibres have an ovoid section (Figs 4.1(A) and 4.2(A)) and the myofibril bundles, which are of regular thickness, occupy most of the sarcoplasm, apart from the space taken up by the mitochondria, which form two columns at the edges of the fibre. In contrast, the architecture of the B fibres is less regular. Their contractile apparatus consists of myofibril bundles that are irregular in section but which have a fairly constant width; large mitochondria lie between the myofibril bundles (Fig 4.1(B) and (C)). The myofibril bundles in both fibre types are of equivalent thickness. They are bordered with membrane invaginations arising at the level of the M-line. In the B fibres (Fig. 4.3) these invaginations are coupled to tubules of the sarcoplasmic reticulum (SR) and hence are morphologically analogous to the transverse tubular (T) system of vertebrate muscle fibres (Duvert and Salat 1980). However, it seems unlikely that such invaginations can have the same role in excitation—contraction coupling as in the vertebrate T-system, because in the A fibres the tubules of the SR are not coupled to the membrane invaginations, but are found adjacent to the surface membrane (Fig. 4.2(B) and (C)). Fig. 4.4 illustrates these different arrangements schematically. The volume density of the SR in both

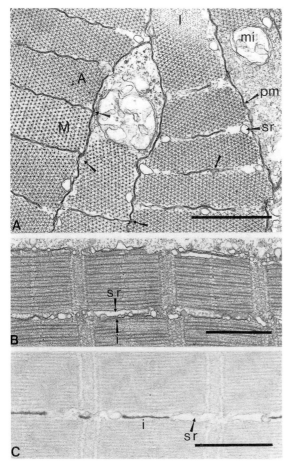

Fig. 4.2 (A) Transverse section of an A fibre group showing the continuity between the invaginations and the plasma membrane (arrows) at the M-level. A, I, M; different sarcomere levels; mi, mitochondrion; pm, plasma membrane; sr, sarcolasmic reticulum. Scale bar: 1 μm. (B) Longitudinal section of B fibre showing internal couplings. Scale bar: 1 μm. (C) Longitudinal section of A fibre from animal soaked in sea water with HRP prior to fixation, showing extracellular spaces of the membrane invaginations (i) containing HRP (dark), and sarcoplasmic reticulum (sr) Scale bar: 1 μm.

fibre types is comparable, although its mode of growth and architecture is different (Dress and Duvert 1983).

Ca^{2+} and Ca^{2+}-ATPase The Ca^{2+} content of the B fibres is higher than that of the A fibres. Histochemical tests reveal its presence throughout the SR, in the mitochondria (especially under 'non-physiological' conditions), and at the level of the surface membrane (which shows a Ca^{2+}-ATPase activity) (Savineau and

Duvert 1986; Duvert *et al.* 1988a). Ca^{2+}-ATPase activity is also seen associated with the Z-lines of the myofilaments, (Duvert *et al.* 1988(b)) but, although quantitatively different in the two fibre types, Ca^{2+}-ATPase does not allow differentiation of ontogenetic stages of the fibres. The significance of this enzymatic activity is unknown, but since it is not present at the A-band level, it does not seem directly concerned with actin–myosin interaction.

The contractile apparatus of the A and B fibres is indeed remarkable. Its structure (Fig. 4.2(A)–(C) and Fig. 4.3) recalls that of arthropods, as was recognized by Camatini and Lanzavecchia (1966) and Halvarson and Afzelius (1969). Amongst other striking features are the very short sarcomeres (1.55 ± 0.2 μm) and A-bands (1.2 ± 0.17 μm). The primary filaments are tubular in section (17.4 ± 1.9 μm in diameter) and are separated from six neighbouring secondary filaments by approximately 26 nm. The ratio of secondary to primary filaments is 3:1. These myofilaments are arranged in an extremely regular double hexagonal array. At the H-level, the primary filaments are triangular in section and have a particular arrangement which raises problems of interpretation (Afzelius 1969). Extensible C filaments link the primary filaments at the Z-line (contrary to earlier reports). The very short I-bands (± 0.3 μm long) can only permit a small degree of shortening from the resting state.

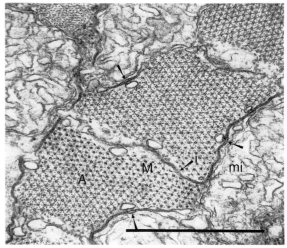

Fig. 4.3 Transverse section of a B fibre showing couplings and thick tubular myofilament array, incomplete at fibre periphery. Mitochondria occupy a large part of fibre volume (compare with Fig. 4.2(B)) Scale bar: 1 μm.

Junctions between fibres

Both types of primary muscle fibres are linked by several different kinds of junctions (Duvert *et al.* 1980*b*).

First, there are numerous gap junctions, which not only occur between fibres of the same type but also between A and B fibres (Fig. 4.5(A) and (B)). A similar arrangement of a sheet of muscle fibres linked by gap junctions is found in the locomotor systems of other invertebrates, but the functional implications of the chaetognath arrangement have not yet been satisfactorily resolved (see p. 42).

Next, there are zonulae adhaerens (Fig. 4.5(C) and (D)) and macular and zonular columnar junctions (Fig. 4.5(B) and (C)), as in epithelial tissues where the cells form a coherent whole. The Z-lines in A and B fibres are wide ($\pm 0.18 \ \mu m$) and are penetrated by a network of cytoskeletal filaments which are related to the columnar junctions uniting adjacent fibres (Fig. 4.5(B)). Thus, not only are the myofibrils within a fibre mechanically linked, but so too are those of neighbouring fibres. From the functional point of view, this organization is comparable to the connective tissue endomysium and the cytoskeleton of muscle fibres in other groups (e.g. vertebrates), and must play a role in the transmission of force within the musculature. Visual obser-

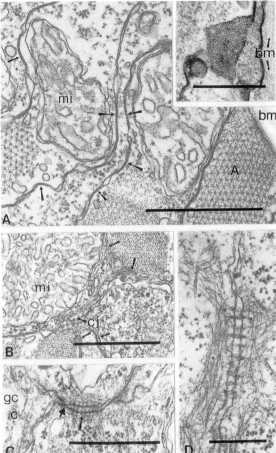

Fig. 4.5 (A) Transverse section of primary musculature showing gap junctions (arrowed) between muscle fibres. mi, mitochondrion; bm, basement membrane. Scale bar: 1 μm. Inset: tangential section of gap junction impregnated with ruthenium red. Scale bar: 0.5 μm. (B) Macular columnar junction (cj) and associated cytoskeleton (arrows) at the I-band level. Scale bar: 1 μm. (C) Zonular columnar junction (c) near body cavity (gc), separated from it by a zonula adhaerens (arrowed). Scale bar: 1 μm. (D) Higher power view of the junction showing relations between the various components. Scale bar: 0.5 μm.

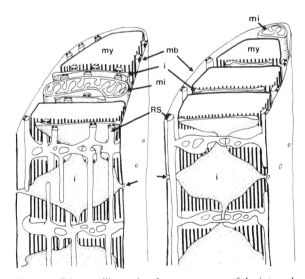

Fig. 4.4 Diagram illustrating the arrangement of the internal tubular systems in the two types of primary muscle fibre (A fibre on right, B on left). The arrows show the connexions between the invaginations (i) and the plasma membrane. mb, plasma membrane; mi, mitochondria; my, myofilament bundle; RS, sarcoplasmic reticulum.

vations of intact animals swimming show that the primary musculature is capable of very rapid, brief contractions (Goto and Yoshida 1981), swimming speed being in excess of 14 cm s⁻¹ (Ducret 1977). However, the swimming movements (accompanied by rotation of the body along its axis (Grassi 1883)) are varied and poorly known (Reeve and Walter 1972*a*; Feigenbaum and Reeve 1977; and Goto and Yoshida

1981). Records of primary muscle activity from intact animals held by suction electrodes and from pinned-out preparations show that during the regular rhythmic bursts of activity they contract at 20–40 Hz, with dorsal and ventral quadrants alternating in activity (Bone, unpublished data). Directional swimming is presumably brought about by differential activity in the different quadrants, possibly being related also to the activity of the secondary musculature (see below).

The ultrastructural features of the primary muscle, together with its mechanical performance, make it similar in a number of respects to the indirect flight muscle of arthropods, a group far removed from chaetognaths. As Halvarson and Afzelius (1969) pointed out, the taxonomic and phylogenetic value of such similarities remain to be established, the more so since the septate junctions reported in *Sagitta* are totally different from the intercellular junctional types in arthropods (Green and Bergquist 1982) and the innervation of the primary muscle fibres is entirely different to that of arthropods.

As Duvert and Barets (1983) showed, the nerve terminals supplying the primary muscle fibres lie in the epidermis, separated from the muscle fibre membrane by the basement membrane (Fig. 4.6).

Secondary musculature

A second, much less abundant, type of muscle (making up < 1 per cent of the trunk wall volume) was dis-

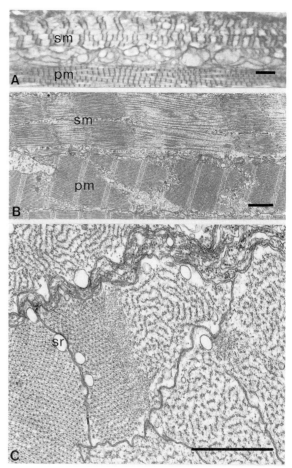

Fig. 4.7 (A) Longitudinal semithin section showing secondary muscle fibres (sm) above, and primary muscle fibres (pm) below. Scale bar: 10 μm. (B) The same at the ultrastructural level. Note alternation of sarcomere types in secondary muscle. Scale bar: 1 μm. (C) Transverse section of the two sarcomere types of the secondary muscle. (sr), sarcoplasmic reticulum; (i) infolding of sarcolemma. Scale bar: 1 μm.

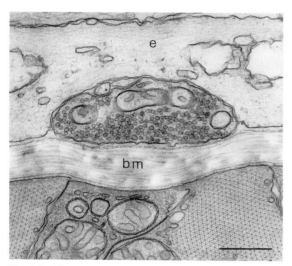

Fig. 4.6 Motor terminal in epidermis (e) separated from muscle fibre by basement membrane (bm). Scale bar: 0.5 μm.

covered by Grassi (1883). This muscle type is strictly localized at the root of the mesenteries, against the basal membrane (Fig. 4.6), forming two lateroventral bands at the edges of the lateral fields (Duvert 1969b; Duvert and Salat 1979). The longitudinal fibres of this muscle type may be organized into bundles by the myoepithelial cells described below. The striking feature of this muscle type (unparalleled, so far as is known, in any other animal) is that the contractile apparatus is made up of two different kinds of sarcomeres which alternate in series (Fig. 4.7). This unique arrangement was observed by Schmidt (1951; 1951–2) who did not understand its significance. At the level of the 'classical' type of sarcomere, the ratio of primary to secondary filaments is 10:1. The other sarcomere type is of variable length and only contains a priori a single major class of filament, reticulated by transverse links and related to an ill-defined structure (Fig. 4.7(B) and (C)). The two types of sarcomere in a single fibre appear able to contract in a relatively independent manner. Other cytological characteristics suggest that this is a 'slow' muscle. Thus, in longitudinal sections, when the primary musculature is contracted, the secondary fibres may be folded with their sarcomeres apparently uncontracted. Because of its position, the secondary musculature may perhaps act (at least in part) to direct the body of the animal as it swims. The myoepithelial cells associated with it have their contractile apparatus disposed perpendicularly to that of the primary and secondary musculature, transversely across the trunk. In view of this orientation, they may act to adjust the pressure within the hydroskeleton, which presumably must vary during ingestion of prey and during the reproductive period. In preparations opened and pinned-out, with the body cavity exposed, slow, intermittent transverse contractions are seen; presumably these are brought about by the myoepithelial cells.

The association of the secondary muscle and myoepithelial cells, which contract at right angles to each other, and (in some forms) the transverse muscles (see below), all of which are separated from the primary musculature, suggests that this division into histological territories in the trunk ('paralleling' a division into separate organs) is required as an adaptation for rapid movements in an animal built on an essentially epithelial plan.

Myoepithelial tissue

Certain cells, linked to the lateral fields, surround groups of secondary muscle fibres. These cells have their contractile regions applied to the basal membrane. The apical pole is secretory and is evidently an important cellular centre, rich in endosomes and with numerous coated vesicles. These features suggest that the interchanges between the fluid of the body cavity and the tissues of the body wall take place via the space between the thin endothelium which lines the cavity and the other tissues. As proposed by Welsch and Storch (1982), rather than being a simple covering, this endothelium probably plays a central role in transport phenomena between the fluid phase of the body cavity and the tissues of the trunk wall, whose paracellular pathways are open to the surrounding sea water. It is in this space that the exchanges with the myoepithelial cells (which have the status of an endocrine gland—a structure hitherto undescribed in chaetognaths) take place.

The other tissues of the trunk wall

These are of several kinds. The least well defined, whose nature is not yet precisely known, are those that make up the 'mesenteries', and the 'lateral fields' (which, with the epidermis, are the only tissues containing glycogen), and the endothelium, which lines the body cavity. Welsch and Storch (1983a) thought that the epithelium was a coelomic lining whose characteristics link chaetognaths with archicoelomates; this has yet to be confirmed. The origin and nature of the endothelium has not been established: it is traversed by cilia arising from the lateral fields and the body cavity itself is a neoformation (Doncaster 1902) rather than a coelom (Chapter 1). The endothelium is linked to the muscular tissue by gap junctions and contains granules (Fig. 4.8). The surface in contact with the fluid of the body cavity is covered with a fibrous coat which is PAS negative and rich in anionic groups.

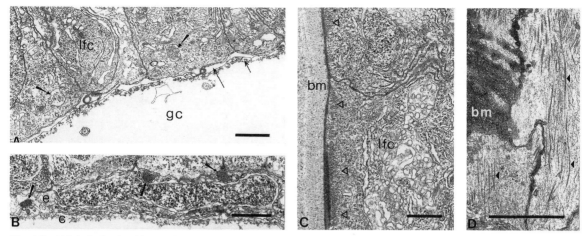

Fig. 4.8 (A) Apical pole of lateral field cells (lfc). Numerous vesiculotubular profiles with coated vesicles are seen in the cells, and also some secretory (?) granules (arrowed). A fibrillogranular coat (c) covers the thin epithelial cells (e) lining the body cavity (gc). Note two cilia in the body cavity. Scale bar: 1 μm. (B) Section of squamous epithelial cell (e) lining body cavity (below). Note granular coating (c) facing body cavity, and cytoplasmic granules (arrowed) above cell nucleus. Scale bar: 1 μm. (C) Vertical transverse section of outer border of lateral field cells (lfc) adjacent to basement membrane (bm), showing myofibrillar apparatus (arrowed) of these myoepithelial cells. Scale bar: 1 μm. (D) Longitudinal tangential section at same level showing the large thick filaments (arrowed) of the myofibrillar apparatus. Scale bar: 1 μm.

Transverse muscles

In several species, which Tokioka (1965c) regarded as the more primitive, there are transverse muscle fibres in the anterior part of the trunk. These run transversely and obliquely from just below the region of the ventral ganglion across the body cavity to the upper edge of the ventral primary muscle quadrant. This arrangement is seen in *Eukrohnia hamata* (Fig. 4.9(A) and (B). Fig. 4.9(C)) shows the ultrastructure of the transverse fibres in *Spadella cephaloptera*. The innervation of the transverse muscles differs from that of the primary muscle fibres because the transverse muscle fibres in *Spadella* cross the basement membrane locally to reach nerve terminals in the epidermis (Duvert, unpublished data); the fibres of the bicornis muscle in the head (Duvert and Barets, 1983) also behave in this way. If Tokioka is correct in supposing that the more primitive species are those possessing transverse muscles, they may perhaps be regarded as vestiges of a more extensive musculature in a benthic ancestral form.

Nothing is known of the function of the transverse muscles.

Visceral musculature

Striated muscle cells occur around the oesophageal region of the gut. Elsewhere the gut is encircled by unstriated smooth muscle cells (*Sagitta setosa*). In *S. elegans* (where the gut cells are largely expanded into sacs containing NH$_4$) muscle cells are absent from all but the most dorsal and most ventral portions of the gut, where they join the mesenteries. The gut cells are ciliated and may assist in food transport, but peristaltic movements have been reported by Burfield (1927), and these are presumably brought about by the smooth muscle cells. Although small visceral nerves run along the top and bottom of the gut (Chapter 6) and small bundles of axons are sometimes seen in the mid-region, innervation of the smooth muscle layer has not been observed and nothing is known of the pharmacology of the cells.

Physiology and biochemistry

Physiological studies have been limited to *S. elegans*, *S. setosa*, and *S. friderici* (the later being physiologically indistinguishable from *S. setosa*) and have concerned the ionic basis of the muscle potentials and the role of external Ca^{2+}, neuromuscular transmission and acetylcholine receptors, some preliminary mechanical experiments, and the existence of electrical coupling between fibres. Biochemical studies on *S. setosa* and *S. friderici*, have concerned enzyme levels and the contractile proteins in the primary muscle.

The ionic basis of the muscle potentials

It is remarkable that two groups of workers have come to quite different conclusions about the ionic basis of muscle potentials.

Schwartz and Stühmer (1984) studied the locomotor muscle fibres of *S. elegans* using loose patch and intracellular electrodes. Rapid action potentials (overshooting some 20 mV from resting potentials of 67 mV) were obtained following extracellular electrical stimulation. Resting potentials of approximately 70 mV were also found in *S. setosa* by Duvert and Savineau (1986) who concluded that they were K^+-dependent. Voltage clamp analysis (Fig. 4.10) showed rapid voltage-dependent inward currents which were largely blocked by the Na^+ channel blocker tetrodotoxin (TTX) and which showed the rapid voltage-dependent inactivation typical of Na^+ channels. Sometimes a slower inward current was observed, this was insensitive to TTX but much reduced by the Ca^{2+}-channel blocker Co^{2+}. Furthermore, vigorous twitches evoked by electrical stimulation of whole animals in sea water were blocked within 60 s in artificial sea water lacking Na^+ (0NaASW); the animals remained flaccid until Na^+ was added, when they again responded to stimulation. Placing the animals in artificial sea water lacking Ca^{2+} (0CaASW) did not abolish contractile responses. These

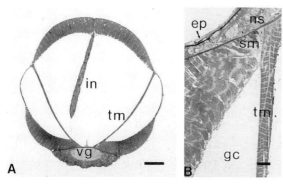

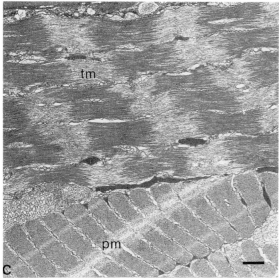

Fig. 4.9 (A) and (B) Transverse semithin sections of *Eukrohnia hamata* showing the transverse muscle (tm). bm, basement membrane; ep, epidermis; gc, body cavity; in, intestine; pm, primary musculature; sm, secondary musculature; ns, nerve bundles; vg, ventral ganglion. Scale bars: 10 μm and 5 μm. (C) Transverse section of *Spadella cephaloptera* showing the transverse muscle fibres (tm) in longitudinal section. Primary muscle fibres (pm) below. Extracellular space is dark in this preparation. Scale bar: 1 μm.

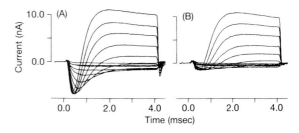

Fig. 4.10 Voltage clamp records from a patch of primary muscle fibre membrane in *Sagitta elegans*. All records begin at holding potential and show current flowing when the membrane was depolarized from 10–120 mV in 10 mV steps. The early voltage-dependent, rapidly inactivating inward current seen in (A) (normal seawater) is largely abolished in (B) after the preparation was exposed to TTX (500 mM). From Schwartz and Stühmer (1984) with permission.

observations on *S. elegans* provide strong evidence for the view that the action potentials are carried mainly by Na^+, and that the Ca^{2+} required for activation on the contractile apparatus is derived from intracellular stores. As noted on page 33, chaetognath muscle fibres have an elaborate internal compartment system open to the external medium, which is in some ways analogous to the vertebrate muscle fibre T-system. In the B fibres, the intracellular sacs of the SR are coupled to this tubular system, whereas in the A fibres, the sacs of the SR are coupled directly to the sarcolemma.

Savineau and Duvert (1986) found that contractions of whole animals, and of electrically-stimulated muscle strips, were completely and rapidly blocked when sea water was replaced by 0CaASW (contractions were restored by the addition of Ca^{2+}) and that hypercalcic solutions (18.4 mM Ca^{2+}) elicited supranormal mechanical twitches (Fig. 4.11). Moreover, contractions were blocked by the calcium blockers La^{3+}, Co^{2+}, and Mn^{2+}, and were restored by washing in sea water. Duvert and Savineau (1986) showed further that TTX (6–10 g/ml) reduced the amplitude of electrically-stimulated mechanical twitches by 50 per cent, but K^+ contractures with this TTX concentration were not reduced as compared with those in normal sea water. Bone *et al.* (1987*a*) confirmed some of these results, showing that electrical stimulation ceased to evoke contractions in 0CaASW, or in sea water containing

Mn^{2+} or Co^{2+}, and that contractions resulting from application of the putative neuromuscular transmitter acetylcholine (ACh) were also reversibly blocked by these calcium channel blockers. Savineau (personal communication) has found that contraction is unaffected by replacement of sea water by artificial sea water containing only 50 per cent of the normal Na^+ level, and that replacement of Na^+ by Li^+ elicits increased resting tension whilst reducing the amplitude of electrically-stimulated mechanical twitches. In *S. setosa*, therefore, the evidence available indicates that external Ca^{2+} is required for contraction and that the contractile apparatus is not activated directly by release of Ca^{2+} from internal stores. Furthermore, it appears that the muscle potentials are carried mainly by Ca^{2+}, Na^+ playing a minor role.

The conflicting results obtained from these two chaetognath species certainly require further investigation. On the one hand, the experiments on *S. setosa* do not exclude the possibility that a small external Ca^{2+} component of the action potential is required to trigger release of Ca^{2+} from internal stores and Savineau and Duvert (1986) provide cytochemical evidence that both the SR and mitochondria might act as such stores, as well as showing the presence of external Ca^{2+} at various external sites around the fibres. In view of the extensive SR system in both A and B fibres (of similar volume density in both types) involvement of intracell-

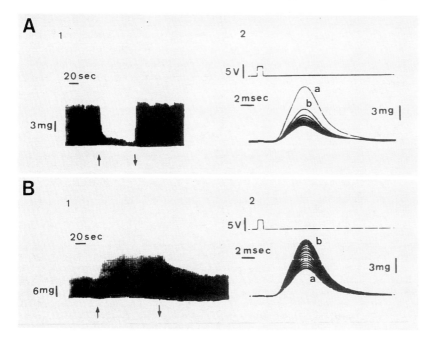

Fig. 4.11 Effects of changes in external Ca^{2+} on mechanical twitches of *Sagitta* primary muscles. Arrows below record show application and removal of test solutions. Stimulation frequency: 5 Hz. (A) The effect of 0CaASW, (B) The effect of hypercalcic ASW (see text). In records on right (2) a, twitch recorded in ASW; b, twitches recorded just after test solutions applied. From Savineau and Duvert (1986).

ular stores seems more probable than that all the Ca^{2+} required to activate the contractile apparatus enters during the action potential. On the other hand, the immediate effect of 0NaASW solutions in blocking contractions in *S. elegans*, and the lack of effect of 0CaASW solutions, suggest that internal Ca^{2+} stores are activated by an Na^+ action potential, without requiring entry of external Ca^{2+} during the action potential. However, Co^{2+} (10 mM) rapidly (and reversibly) blocks contractions of intact *S. setosa* and *S. elegans* in response to mechanical stimuli (Bone, unpublished data), suggesting that external Ca^{2+} is required in both species. It is difficult to believe that

the two species could differ in the ionic basis of their action potentials, although if this were the case it would be interesting indeed: further work (for example with caffeine to deplete intracellular Ca^{2+} stores, and voltage clamp experiments in 0Na solutions) is needed to resolve the present dichotomy of the experimental results.

Neuromuscular transmission

In contrast to the unsatisfactory situation with the ionic basis of the action potential, there is agreement that ACh is likely to be the main neuromuscular transmit-

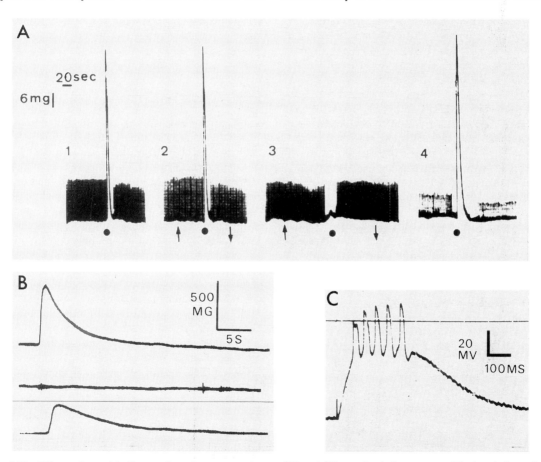

Fig. 4.12 Effects of acetylcholine on the primary musculature. (A) and (B), mechanical responses, (C) electrical membrane responses. (A) Effects of 5 s applications (at dot below record) of acetylcholine (10^{-5} M) to muscle strip. Other drugs applied and removed at arrows below record. 1, acetylcholine (ACh) alone; 2, ACh in the presence of 10^{-4} M atropine; 3, ACh in the presence of 10^{-5} M *d*-tubocurarine; 4, ACh in the presence of 10^{-6} g ml^{-1} TTX. From Duvert and Savineau (1986). (B) Entire animal. The upper record shows the response to acetylcholine added to the bath, the middle record shows that this response is abolished by *d*-tubocurarine (10^{-4} M), and the bottom record shows that the response is restored by washing with sea water. From Bone *et al.* 1987*a*. (C) Electrical membrane responses from primary muscle fibres following iontophoresis of acetylcholine. From Bone *et al.* 1987*a*.

ter. Duvert and Savineau (1986) found that ACh (10^{-5} M) evoked strong contractions of the locomotor muscle (Fig. 4.12) blocked by *d*-tubocurarine (10^{-5} M), and that the ACh response was not changed by the addition of TTX (10^{-6} g/ml). Since atropine had no effect on the cholinergic response, these authors suggested that the receptors were nicotinic. Other 'classical' neurotransmitters tested, such as dopamine, adrenalin, and noradrenalin had no effect and gamma-amino butyric acid (GABA), glutamate, and serotonin (5-HT) gave small reductions in the amplitude of electrically-evoked twitches. These results were confirmed by Bone *et al.* (1987*a*), who showed in addition that iontophoretically-applied ACh evoked depolarizations and spikes similar to those seen during the characteristics rhythmic 'spontaneous' activity of pinned-out preparations (Fig. 4.12). In line with these physiological observations, acetylcholinesterase is histochemically demonstrable between the muscular and the epidermal layers (where the motor endings are found). Most recently, Grandier-Vazeille *et al.* (1989) have detected (by the use of auto-anti-idiotypes (AAIs) raised against an ACh conjugate) ACh binding sites that seem to be associated with the muscle membrane in different head muscles. Although the locomotor trunk musculature was not examined, this result supports the view that ACh is the neuromuscular transmitter. *Spadella* also contracts in response to ACh, but a number of other criteria obviously have to be satisfied before definitely concluding that ACh is the chaetognath neuromuscular transmitter and this is not to say that other neurotrans-

mitter substances may not be present at the neuromuscular junctions. It is certainly possible that some neuropeptide 'modulation' of cholinergic responses may occur. Duvert and Barets (1983) observed some dense-cored vesicles amongst the electron-lucent vesicles of the nerve terminals assumed to innervate the locomotor musculature and Goto and Yoshida (1987) figure a similar terminal with much more abundant dense-cored vesicles. Again, immunocytochemistry has shown the presence of several neuropeptides in the widespread epidermal plexus and these may, in part, be 'motor' to the locomotor musculature (they include glutamate, aspartate, galanine, beta-endorphin, and RFNH$_2$-like substances). The ease with which access can be obtained to the locomotor muscle fibres and the variety of possible 'modulators' observed, suggests that there are attractive possibilities for further experimentation on the effects of cholinergic modulating agents.

Mechanical properties

Preliminary mechanical studies of the trunk muscle fibres (Duvert and Savineau 1986) have shown them to contract very rapidly. Electrical stimulation with 0.5–1.0 ms current pulses evokes brief isometric mechanical twitches lasting some 18 ms, time to peak tension being around 5 ms (Fig. 4.11). Electrical stimulation of strips of trunk muscle produced maximum tensions of approximately 1.3×10^{-4} N, whilst addition of ACh to the muscle bath produced much greater tensions, of approximately 7.4×10^{-4} N. Addition of ACh to a bath containing an entire animal (Fig. 4.12(B)) produced tensions of 6.5×10^{-3} N. Tetanus occurred at stimulation frequencies above 50 Hz.

Whilst these preliminary observations give some idea of the extreme rapidity of the system, as yet nothing is known either of the forces exerted per unit of cross-sectional area, or of the force/velocity relationship. Nor is anything known of the modes of operation of the A and B fibres, and what part each plays in the mechanical records shown above. This is perhaps the most obvious area for further research particularly in view of the observations reported in the next section, and the unique structure of the secondary muscle fibres.

The problem of electrical coupling between muscle fibres

There is clear evidence from the ultrastructural studies described in the first section of this chapter that there

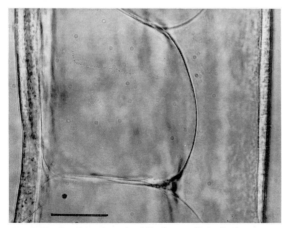

Fig. 4.13 Portion of trunk of *S. elegans* (living), showing gut sacs in the body cavity. Scale bar: 100 μm. From Bone *et al.* 1987*b*.

are two main types of muscle fibre in the primary locomotor musculature (at different stages of development) and that they are linked to adjacent fibres by gap junctions. (They are linked not only to fibres of the same type, but also to fibres of the other type). Gap junctions normally provide low resistance electrical pathways between the cells that they couple, such that electrical events across the cell membrane of one cell are seen in the adjacent cell; in the case of coupled muscle cells this means that if one is activated so too is its neighbour. As yet, the innervation pattern of the locomotor muscle fibres is not definitely known, although it appears in many instances that a single nerve terminal may innervate several muscle fibres, and that each muscle fibre receives at least one nerve terminal.

Thus it is far from obvious why there are two different fibre types, and why their innervation should be so widespread. Earlier attempts to demonstrate electrical coupling between muscle fibres in *S. setosa* by current injection and recording from an adjacent fibre failed (Bone *et al.* 1987a). More recently, similar experiments and Lucifer yellow injections have, in some cases, shown both electrical coupling and dye coupling between small groups or bundles of fibres, although coupling does not appear to exist between different groups of coupled fibre bundles. In other cases, dye injection shows single fibres that are not dye-coupled to adjacent fibres, and it was presumably from such fibres that no electrical coupling to other fibres was earlier found. Evidently, further investigations of this curious situation are needed, and it should be borne in mind that the coupling of fibres by gap junctions may reflect metabolic 'coupling' in an animal lacking a circulatory system and where the B fibres of

the primary musculature have no direct link with the boundary of the body cavity.

Muscular activity can certainly be graded in chaetognaths; they can make relatively slow, small movements as well as very rapid, large flicks of the body, but how this variety of movement is brought about is unknown. Evidently, it could result from differential activation of different muscle fibre types, or even recruitment of some kind of motor unit, perhaps represented by the bundles of coupled fibres. Unfortunately, nothing is known of the number of muscle fibres innervated by a single ventral ganglion motoneuron.

Primary muscle enzymes

Apart from the work on Ca^2-activated myosin ATPase mentioned on page 34, a preliminary study of enzyme activity levels in homogenized tissue (Storey, personal communication) showed that both L- and D-lactate dehydrogenase activities were below 0.02 μmol substrate min^{-1} g wet wt.$^{-1}$, whilst malate dehydrogenase activity was 1.9 μmol min^{-1} g wet wt.$^{-1}$. There is little lipid in chaetognaths, and further enzymatic work is evidently needed to establish the basis of muscle metabolism.

Contractile proteins

Using SDS-polyacrylamide gel electrophoresis and SDS-PAGE and 'Western' immunoblotting, myosin, paramyosin, actin, and caldesmon (an actin-and calmodulin-binding protein) have been identified in crude extracts of *Sagitta* muscle (Grandier-Vazeille and Duvert 1989). Other contractile proteins are also present but have not yet been definitely identified.

Buoyancy

Although very few estimates have been made of the densities of different species, most chaetognaths are probably slightly more dense than the sea water in which they live (Kuhl (1938) observed that *S. setosa* was slightly more dense than sea water and Feigenbaum and Reeve (1977) found that *S. hispida* was denser than sea water) and are therefore, obliged to swim to maintain their horizontal position in the water column. There are, however, a few species in which there are special buoyancy adaptations, and these species are probably close to neutral buoyancy in their

normal environment. It seems likely that further work may extend the number of species of this latter kind, and also increase the types of buoyancy adaptations found in chaetognaths.

Since chaetognaths are constructed from materials that are denser than sea water (in particular, the muscle proteins and chitinous hooks and teeth), two strategies are possible for reducing density; storage of materials that are less dense than sea water, or reduction in dense components. Both seem to be used by different species. Investigation of *S. elegans* has shown that this species stores NH_4^+ ions in the vacuoles of greatly enlarged gut cells (Bone *et al.* 1987*a*)). These gut cell sacs (Fig. 4.13) almost completely fill the body cavity, (which in *S. setosa* is filled with a fluid similar to sea water). The difference in density between Na^+ and the lighter NH_4^+ ions accounts for the lower density of *S. elegans*. Presumably the fluid in the gut sacs of *S. elegans* is isosmotic with sea water, in some sacs it contained 500 mM of NH_4^+, although mean levels were approximately 250 mM, sufficient to make the animal close to neutral buoyancy. As Dallot (1970) has shown, several other chaetognath species possess gut sacs similar to *S. elegans* and, although the fluid in their sacs has not been analysed, it is likely that they also may store NH_4^+ ions in the same way.

Other chaetognath species seem to use a different system of storage of light materials, although this awaits experimental confirmation. In *S. lyra*, and *S. planctonis*, Kapp (personal communication) has found that the fins are swollen with gelatinous material, which is assumed to be less dense than sea water. *S. lyra*, *S. hexaptera*, *S. enflata* and the curious *Bathybelos typhlops* show a marked reduction in the musculature (the chief dense component of the trunk), which is apparently the other strategy for reducing density seen in chaetognaths. Although the density of such species has not been determined experimentally, Feigenbaum (1977) observed that *S. enflata* appeared to be neutrally buoyant in the laboratory.

The significance of neutral buoyancy or density reduction in chaetognaths has remained unclear, although it seems evident that it must be related *inter alia* to the techniques used in capturing prey; searching by the hop-and-sink method in dense species, and ambush predation by neutrally buoyant species (Chapter 5). However, there seems to be no significant difference between the rhythmic activity of *S. elegans* and *S. setosa* in the laboratory, and these species are of very different densities.

Although accurate density determinations of chaetognaths are difficult to make because the animals are so small, it is practicable to make *relative* density measurements, even aboard ship (provided the density of the sea water in which measurements are made is known), as Kapp (personal communication) has pointed out, and it would be well worthwhile examining those species where low density gelatinous materials have been postulated, or where the musculature is reduced.

5

FOOD AND FEEDING BEHAVIOUR
David Feigenbaum

Introduction

All species of chaetognaths are carnivorous and may feed on several trophic levels. The benthic species are small and are minor constituents of their ecosystems but planktonic chaetognaths are often abundant. Since they are food for a wide variety of larger organisms (Bigelow 1924; David 1955; Reeve 1966) they occupy a central and sometimes important position in planktonic food webs.

Early reports of the phylum provide little information about feeding other than diet. Chaetognaths are delicate and, until recently, were difficult to collect undamaged and maintain in the laboratory. However, major advances have been made in recent years. Today, it is possible to estimate the daily ration and food web significance of almost any species. A summary of the methods and results are presented in this chapter. A more detailed discussion is presented in Feigenbaum and Maris (1984). The reader is advised to consult this review before embarking on a feeding study.

The feeding process

Chaetognaths detect prey by sensing movement. In the laboratory, attacks have been experimentally induced against low frequency vibrating probes placed within millimeters of the animal's body (Horridge and Boulton 1967; Feigenbaum and Reeve 1977, see Chapter 3). Small organisms such as copepods, barnacle nauplii and appendicularians, which make a distinct beat while moving, appear to be the most susceptible to attack. Fish larvae and other chaetognaths may be detected by their tail beat, although lateral motion alone may be sufficient (Feigenbaum and Reeve 1977).

Planktonic chaetognaths initiate an attack with a sudden flex and flick of their tail. Benthic species, which are generally adherant to the substrate, make a rapid upward jerk of the forward part of their body when prey swims overhead. The hold on the substrate is maintained. In both chaetognath groups, attacks cover short distances; the prey is not pursued if it is missed or struggles free (Parry 1944; Feigenbaum, personal observation).

Prey is grabbed by the chaetognath's hooks ('grasping spines') in a quick action during which the hood is folded backward and the mouth projected forward. The chitinous hooks act like rigid fingers, moving individually to manipulate and stuff the prey down the chaetognath's 'throat' (Parry 1944; Feigenbaum, personal observation). In *Spadella schizoptera* the lateral plates and musculature associated with the hooks take on the appearance of a forearm when the chaetognath is feeding. Prey are rarely pierced by the hooks, yet seem to be immobilized before they are ingested (Nagasawa 1985c; Feigenbaum, personal observation). Thuesen (Chapter 6) has shown that the Na$^+$ channel blocker, tetrodotoxin (TTX), is used to immobilize the prey.

Rigid prey is manipulated by the hooks and eaten endwise, sticking out of the mouth for long periods of time in some cases. Soft-bodied prey such as fish larvae are sometimes folded over (Kuhlmann 1977). Thus the orientation of prey in the gut cannot be used to indicate whether the prey was moving towards or away from the chaetognath at the time of capture.

Prey size is limited to organisms which are large enough to be handled by the hooks and small enough, in at least one dimension, to pass through the mouth. On rare occasions large prey become lodged in the forward gut and the chaetognath dies (Nagasawa 1985c).

The role of the one or two rows of small teeth is still

conjectural. It is unlikely that they are used to grasp or hold prey, as some have suggested (John 1933; Parry 1944; Hyman 1959; Furnestin 1977) because prey are easily controlled by the hooks. Furthermore, some genera lack teeth. The teeth are not hollow (Cosper and Reeve 1970; Bieri *et al.* 1983) but several investigators have nevertheless suggested that they are used to inject a toxin into prey, to paralyse them and prevent damage to the chaetognath gut lining (Burfield 1927; Bieri *et al.* 1983). However, Nagasawa (1985c) could not confirm this hypothesis in her scanning electron microscope study of captured copepod exoskeletons.

Once ingested, prey are wrapped in a peritrophic membrane (Reeve *et al.* 1975; Sullivan 1977) and passed to the posterior gut by peristaltic movements of the gut wall (Parry 1944). This process may take from a few seconds to more than 30 minutes, depending on species and temperature (Reeve *et al.* 1975; Feigenbaum 1982). Once in the posterior gut, prey may be moved back and forth throughout the gut or may remain near the anus until defaecation. Observations of both have been made, sometimes for the same species (Grey 1930; Reeve *et al.* 1975; Canino and Grant 1985; Feigenbaum, personal observation). The peritrophic membrane may be thick (*Sagitta hispida*) or thin (*S. enflata*). Sheader and Evans (1975) believe that its function is to protect the gut lining from the sharp spines of crustaceans.

After *Spadella cephaloptera* has swallowed a copepod it gulps water, which may serve to carry enzymes liberated by the oesophagus to the ingested prey (Parry 1944).

Many prey can be recognized, and often identified to species, during the digestive process because some parts of their anatomy resist digestion. These parts include the hooks and lateral plates of prey chaetognaths, the mouth parts of crustaceans and the faecal pellets of appendicularians (Sullivan 1977; Feigenbaum 1982; Nagasawa and Marumo 1979b).

After digestion, the prey is defaecated as a soft pellet. If several prey are in the gut they are usually defaecated together, even if some were ingested considerably later than the others (Reeve *et al.* 1975; Nagasawa and Marumo 1984b; Nagasawa 1985c; Feigenbaum personal observation). Digestive efficiency for *Sagitta hispida*, feeding on at least three copepods at a time, averaged 80 per cent (Reeve and Cosper 1975). This is similar to efficiencies reported for *S. crassa* (83 per cent) (Nagasawa 1985c) and typical of carnivores (Conover 1978).

Usually, however, chaetognaths contain one prey at a time (Mironov 1960; Newbury 1978; Feigenbaum 1979a). For example, of 12 171 *S. crassa* examined from Tokyo Bay, 28.3 per cent contained one prey, 4.1 per cent contained two and 0.3 per cent contained three, four or five (Nagasawa and Marumo 1984b). In some instances the frequency of multiple prey can be predicted by a random distribution. In others it cannot, indicating that either the prey distributions are patchy or chaetognath movement patterns are not random (Sullivan 1977). Given the opportunity, hungry chaetognaths will eat several prey in rapid succession (Cosper and Reeve 1975; Feigenbaum, personal observation).

Diet

Chaetognaths begin feeding a few days after hatching; they feed on proportionately small prey. Copepod nauplii are presumably the mainstay of the young chaetognath's diet, but tintinnids, and sometimes rotifers and small meroplankters, are also eaten (Mironov 1960; Reeve 1966, 1970b, 1980; McLaren, 1969; Pearre 1981). Diatoms and dinoflagellates have also been found in chaetognath guts but it is possible that these, and especially the diatoms, may have been inadvertently ingested or derived from the guts of herbivorous prey. Nagasawa (1988) found that diatoms were excreted without faecal pellet membranes, which suggested they were not ingested as food. Many organisms in the plankton are small enough for young chaetog-

naths, but may be underrepresented in the diet because they are difficult for the animal to detect because of their smooth ciliary motion. Reeve and Walter (1972a) found that *S. hispida* would not grow on a diet of ciliates, even though the prey were large enough for capture and handling. Chaetognaths also do not attack fish eggs (Kuhlmann 1977) or *Artemia* eggs, even when swirled in the water (Reeve 1966). In the laboratory, chaetognaths can be induced to attack objects pressed against their side and the rare instances of chaetognaths having consumed inert objects (Carpenter *et al.* 1972) may arise from actual contact. The consumption of tintinnids may be overestimated by researchers because the lorica is refractory and readily identified in the chaetognath gut, while naked prey may be unrecognized and underestimated.

Older planktonic chaetognaths rely principally on copepodites and adult copepods because of the abundance of these prey in the ecosystems of the water column. (Øresland 1987). A small number of chaetognaths are routinely eaten. Barnacle nauplii are frequent prey in coastal waters, as are appendicularians, when present. Cladocerans, euphausiids and meroplankters can be a signficant part of the diet in a given locality (for examples of the chaetognath diet see David 1955; Mironov 1960; Rakusa-Suszczewski 1969; Stone 1969; Nagasawa and Marumo 1972; Pearre 1973, 1974, 1976a; Sullivan 1977; Feigenbaum 1979a, 1982; Kimmerer 1984; Canino and Grant 1985; Nagasawa 1988). Despite numerous reports in the literature (Lebour 1922, 1923; Bigelow 1924; Kuhlmann 1977), fish larvae are infrequently found in chaetognath guts because they are scarce members of the zooplankton.

Erroneous reports of chaetognaths attacking salps, medusae and other prey too large to pass into the mouth (Massuti-Oliver 1954; Hyman 1959; Alvarino 1962; Croce 1963) are artifacts of collecting and preserving methods (David 1955; Reeve 1970a; Feigenbaum, personal observation).

The diets of benthic chaetognaths are poorly known. In the laboratory, *Spadella* can be maintained on natural zooplankton (Feigenbaum 1976, 1977). Presumably these animals normally depend on benthic crustaceans (John 1933).

Since chaetognaths can utilize prey of a wide range of sizes and developmental stages, the actual diet can be expected to reflect the relative frequency of prey abundance and will vary seasonally (Pearre 1976a; Sullivan, 1980; Øresland 1987; compare with Piyakarnchana 1965; Pearre, 1974 and 1976a; Szyper 1978).

Cannibalism

Chaetognaths prey on other chaetognaths of their own, and other, species (Fig. 5.1). By number, chaetognath prey usually forms a small part of the diet. However, by weight the contribution can be significant, particularly for larger size classes, because they feed on larger prey when available (Mironov 1960; Feigenbaum 1979a).

Some species are evidently more cannibalistic then others. Of eight quantitative reports of diets for *Sagitta enflata* found in the literature, cannibalism averaged 9.5 per cent (0.7–44.8) versus only 1.7 per cent (0–3.1) for eight reports of *S. elegans*. The benthic *Spadella* is

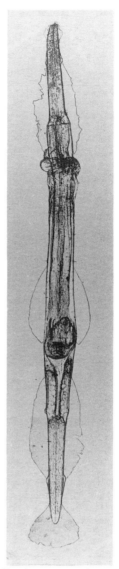

Fig. 5.1 Cannibalism in progress. A composite photograph from Reeve (1971).

rarely cannibalistic (Reeve and Walter 1972b; Feigenbaum personal observation). Although Pearre (1982) proposed that the proportion of chaetognaths in the diet could be predicted on the basis of head width and abundance alone, species differences should also be taken into account.

Reeve and Walter (1972b) suggested that cannibalism is behaviourly related to copulation among mature *S. hispida*. Cannibalism occurs in immature chaetognaths as well. Pearre (1982) pointed out that cannibalism had potential energetic benefits and might be necessary for the existence of large species. If so, the benefits are complex because cannibalism can account for a sizeable part of population mortality in some ecosystems (Szyper 1978; Øresland 1987).

Prey selection

Chaetognaths show selectivity in their feeding behaviour and, as a result, influence the distribution of prey populations. Fulton (1984) illustrated this potential with a series of experimental manipulations. In the experiments, predation by *S. hispida* resulted in a significant decline in abundance of the larger, active copepod, *Acartia tonsa*, and an increase in abundance of the small, less active copepod *Oithona colcarva*.

Although chaetognaths depend on prey movement to detect food, they respond to a broad range of low frequency vibrations (see Chapter 3) and are unlikely to select specific prey species on this basis (Feigenbaum 1977; Feigenbaum and Reeve 1977; Feigenbaum personal observation). Instead, selection may be based on size, or may result from differential movement patterns or escapability of prey. Large chaetognaths eat larger prey than small individuals of the same species (Reeve 1966; Rakusa-Suszczewski 1969; Reeve and Walter 1972a; Szyper 1976; Drits 1981; Feigenbaum 1982; Nagasawa and Marumo 1984a; Øresland 1987). This will often result in a shift in prey species as the chaetognath grows. However, there is a difference in size preference among chaetognath species (Pearre 1980a).

Electivity indices obtained from natural gut contents must be interpreted cautiously because many chaetognaths undergo vertical migrations and may not have fed in the layer in which they were caught. Pearre (1973) used electivities of *S. elegans* to show that the chaetognath fed on the surface and then transported the food downwards. Even where they have been applied to non-migrating stages or species, electivity indices are variable, reflecting the difficulty of estimating the actual availability of prey with a tow net on a scale important to the predator (Sullivan 1980).

Feeding rates

Natural feeding rates can be estimated from an analysis of gut contents if an estimate of the digestion time is also available. The appropriate equations are:

$$FR_n = \frac{NPC\,(24)}{DT}$$

$$\text{(5.1)}$$

$$FR_w = (FR_n)\,(MPW) \qquad \text{(5.2)}$$

where DT = digestion time in hours; FR_n = daily feeding rate in number of prey per day; FR_w = daily feeding rate by weight; MPW = mean prey weight; NPC = number of prey per chaetognath, from gut content analysis.

It is easy to count and measure the prey in a chaetognath's gut because the animal swallows its prey whole and its body is fairly transparent. Quantitative analysis of gut contents makes the following assumptions:

1. Prey species are equally identifiable in the gut.

2. All prey are digested at the same rate.

3. Cod-end feeding is either negligible or can be accounted for in subsequent analysis.

4. The chaetognath neither regurgitates nor defaecates prey during the handling and preservation process.

Assumption 1. is probably false (Pearre 1974); however, this will have little effect on eqn 5.1 as long as the prey can be identified as such, and is not confused with parasites or organisms that may have been inadvertently ingested (e.g. diatoms). Data is also accumulating to prove assumption 2. false (Pearre 1974; Øresland 1987), although the assumption can be accounted for in the same way that 'mean prey weight' accounts for differences in prey sizes. Cod-end feeding (assumption 3) can be handled in a number of ways—principally by eliminating from the analysis food that is forward in the gut or by making adjustments in the collection methods; a detailed discussion is provided in Feigenbaum and Maris (1984). Assumption 4. appears to be valid within acceptable limits. Sullivan (1980) found no loss during preservation in her study. Feigenbaum (1977) used cold formalin to minimize preservation losses in his.

Digestion time (DT) in eqn 5.1 is the time from the ingestion of prey until its defaecation. The most straightforward method of estimating DT is to observe and time feeding in the laboratory. This is also the best method because presumably only healthy chaetognaths will feed in the laboratory. Barring laboratory feeding, DT can still be estimated by placing those chaetognaths collected in the field into an aquarium and recording the times to defaecation of all individuals that appear undamaged. The estimate can then be based on:

1. the longest time recorded;

2. an average of the 'x' longest times (I suggest 5);

3. twice the mean of the times recorded.

A 4th method, used by Szyper (1978), involves preserving groups of chaetognaths at different intervals after capture and then regressing the proportion with prey (p) on the 'time to preservation.' The p = 0 intercept of the regression line provides the estimate. However, this is a cumbersome procedure with several drawbacks.

Table 5.1 lists published digestion times. They range from 40 to 614 min and being a metabolic process, DT varies with temperature (Mironov 1960; Pearre 1981; compare Kuhlmann 1977 with Feigenbaum 1982) and there is some evidence that it also varies with prey size (Nagasawa and Marumo 1984*a*; Øresland 1987). Multiple prey tend to increase digestion time and make it more variable (Reeve 1980; Canino and Grant 1985). The relationship of DT to chaetognath size is unclear. Szyper (1978) and Sullivan (1980) found DT to be unrelated to size. Szyper hypothesized that this was because large chaetognaths digest prey faster than small ones, but eat larger prey that take longer to digest. However, Reeve (1980) found that DT increased with chaetognath size and Feigenbaum (1982) was unable to find a significant relationship

between DT and the ratio of chaetognath weight to prey weight. Individual DT tends to be quite variable, which may be responsible for the lack of clear relationships with other variables. DT may also vary with different populations of the same species. For example, Szyper (1978) estimated DT for the *S. enflata* population of Kaneohe Bay, Hawaii to be 60 min, while Feigenbaum (1979) arrived at 190 min for the Gulf Stream population of this species. Temperatures were similar in both studies.

Chaetognaths are diurnal migrators and often pass through various temperature regimes while feeding and/or digesting prey. As a further complication, the migratory behaviour of young individuals differs from that of older chaetognaths. For accuracy, DT should be based on a mean temperature regime experienced by each chaetognath stage of a given study (Pearre 1981).

Feeding rates of eqns 5.1 and 5.2 are also referred to as the daily ration. Feigenbaum (1979) used these equations to obtain the daily ration of *S. enflata* in the Gulf Stream by chaetognath size. In his study, which excluded young chaetognaths, mean prey weight did not increase with chaetognath size. However both FR_n and FR_w increased because NPC increased with size.

The food containing ratio (FCR) is the percentage of chaetognaths with food in their gut. When multiple prey are rare FCR is sometimes used to estimate NPC. Some studies have used NPC or FCRs to compare feeding rates. This can be satisfactory within a population, providing the different chaetognath sizes remain within similar temperature regimes. However, serious errors can be introduced if the comparison is between separate populations or different species.

Table 5.2 lists feeding rates available in the literature. Daily rations vary from 0.18 prey/day for *S. elegans* in Gullmarsfjorden in late fall, to 60 *Artemia* nauplii/day for adult *S. hispida* feeding in the laboratory at 24°C.

Temperature affects feeding rates. For example, Reeve (1966) found laboratory feeding in *S. hispida* increased with temperature from 10°C to 25°C, but was lower at 30°C. At 33°C the animals died.

The specific daily ration (SDR) is the weight of food consumed daily per unit weight of chaetognath. For convenience SDRs are often based on dry weights. However, they are more meaningful when reported on a carbon, nitrogen, phosphorus or ash-free dry weight basis, because dry weights themselves contain a considerable amount of inorganic salt. Carbon SDRs have been found to be higher than dry weight SDRs in

Table 5.1   Chaetognath digestion times

Species	Temp (°C)	Prey	Digestion time (min)	Reference
Pterosagitta draco	–	Copepods	165	Terazaki (unpub.) cited in Nagasawa and Marumo (1972)
Sagitta crassa	–	*Tigriopsis japanicus*	360	Takano (1971) cited in Nagasawa and Marumo (1972)
	–	*Oithona aruensis*	60	Nagasawa and Marumo
	–	*Acartia clausi*	120–180	(1984*b*)
	22–24	*A. clausi*	102	Nagasawa (1985*c*)
	22–24	*T. japanicus*	168	Nagasawa (1985*c*)
Sagitta elegans	15	Copepods	147	Kuhlmann (1976)
	–	–	210	Reeve (unpub.) cited in Reeve (1980)
	–	–	240	Sullivan (unpub.) cited in Reeve (1980)
	0	Copepods	614	Feigenbaum (1982)
	6	*Calanus*	560	Øresland (1987)
	6	Small copepods	294	Øresland (1987)
Sagitta enflata	40	Chaetognath	40	Grey (1931)
	24–26	Natural zooplankton	60	Szyper (1978)
	23	Copepods	190	Feigenbaum (1979*a*)
Sagitta hispida	–	Copepods	180–240	Cosper and Reeve (1975)
	21	Natural zooplankton	60–120	Reeve (1980)
	21	Natural zooplankton	150–240	Reeve (1980)
	25	Copepods	53	Canino and Grant (1985)
	21	Two copepods	137	Canino and Grant (1985)
Sagitta nagae	–	Copepods	120–300	Terazaki (unpub.) cited in Nagasawa and Marumo (1972)
Sagitta setosa	–	–	300	Parry (1944)
	11.5	–	120	Mironov (1960)
	15.5	–	90	Mironov (1960)
	20	–	60	Mironov (1960)
Sagitta tenuis	–	*S. tenuis* (6.8 mm)	146	Canino and Grant (1985)
	21, 25	*A. tonsa*	69	Canino and Grant (1985)
Spadella cephaloptera	–	–	180–240	Parry (1944)

Sagitta tenuis (Canino and Grant 1985) and *S. enflata* (Feigenbaum 1979*a*).

SDRs are typically high (> 1) for young chaetognaths and fall exponentially with increasing chaetognath size (Feigenbaum 1982; Kimmerer 1984). The more flaccid chaetognath species, such as *S. enflata*, appear less active than rigid species like *S. hispida*, *S. tenuis* and *S. crassa* and have lower carbon based SDRs. The largest *S. enflata* had SDRs of 0.144 (Feigenbaum 1979) and 0.105 (Reeve 1980) compared with 0.220 for *S. tenuis* (Canino and Grant 1985) and 0.40 for *S. hispida* (Reeve 1980). *S. elegans*, with a body type intermediate to these species, had a carbon based SDR of only 0.035 in laboratory feeding (Reeve 1980), but 0.19 in the CEPEX bags (Sullivan unpublished data, cited in Reeve 1980).

Kimmerer (1984) and Szyper (1976) estimated nitrogen and phosphorus (Szyper only) based SDRs for the *S. enflata* population of Kaneohe Bay, Hawaii. The rates are higher than the carbon based SDRs reported by Feigenbaum (1979*a*) for the Gulf Stream population of this species. However, because of the size disparity of individuals between these populations, it is not possible to attribute the difference solely to the type of analysis used.

Environmental factors must also be considered. Feeding in expatriate populations may be affected by stress because the animals have been carried into marginal environments. In a laboratory study, Reeve (1966) found that *S. hispida* had increased feeding rates at the extremities of its salinity range, presumably because stress increased energy demands. In contrast, an expatriate population of *S. enflata* in the surface waters off coastal Virginia fed at half the rate of the population near Miami, Florida (Bushing and Feigenbaum 1984). None of the individuals caught below the thermocline in Virginia had been feeding at all.

Feeding and prey density

In the laboratory, chaetognaths increase their feeding rates with food concentration until they apparently attain satiation (Reeve 1964, 1980; Nagasawa 1984). Reeve called the food level at satiation the 'critical density' and cited its existence in chaetognaths as evidence that they are not superfluous feeders. At food levels above the critical density Reeve found evidence of feeding inhibition.

Attempts to correlate chaetognath feeding in nature with prey abundance have generally been unsuccessful (Mironov 1960; Nagasawa and Marumo 1972; Sullivan 1980 for *S. elegans*; Bushing and Feigenbaum 1984) because chaetognaths do not necessarily feed at the depth at which they are caught (Pearre 1973) and it is difficult to estimate prey density on a scale important to the chaetognaths (Sullivan 1980). Drits (1981) found an increase in the daily ration of *S. enflata* from 0.65 to 3.2 prey/day when copepod density increased from 0.7/1 to 12/1. However, the results were not checked statistically. Kimmerer (1984) found that the feeding of *S. enflata* increased with prey abundance in Kaneohe Bay, but levelled off at a copepod density of 200/1.

In most cases, critical densities determined in the laboratory lie far beyond the range of prey densities in nature, yet some reported natural feeding rates are comparable with maximum laboratory rates (Feigenbaum and Maris 1984). This paradox has been cited as evidence that chaetognaths depend on patches of prey for survival. However, the small percentage of chaetognaths found with multiple prey in nature argues against patch feeding on a regular basis. It is more likely that the discrepancy in required prey densities is due to a feeding inhibition of wild caught animals under laboratory conditions.

Feeding and time of day

Chaetognaths display a diurnal feeding rhythm, with a sizeable increase during the night. This has been demonstrated in the laboratory (Parry 1944; Reeve 1964; Nagasawa 1985*c*) and supported in field studies by higher FCRs and NPCs (Rakusa-Suszczewski 1969; Nagasawa and Marumo 1972; Pearre 1973; Newbury 1978; Szyper 1978; Feigenbaum 1979*a*, 1982; Bushing and Feigenbaum 1984; Kimmerer 1984; Nagasawa 1985*c*). However, due to variability in the data, the differences between day and night levels are often not

Table 5.2 Chaetognath feeding rates

Species	Location	Temperature (°C)	Chaetognath length (mm)	Daily ration No./day	Dry wt basis	Carbon basis	Nitrogen basis	Phosphorus basis	Reference
						Specific Daily Ration			
Pterosagitta draco	Near Hawaii	12–25	5.0–7.0	1.0	–	–	0.02	–	Newbury (1978)
Sagitta elegans	Laboratory Vineyard Sound, Massachusetts	13	16	4[a]	–	0.003–0.035[b]	–	–	Reeve (1980) Feigenbaum (1982)
		0	3.5–20.5	0.53–1.33	0.465–0.006 (3.5mm–20.5mm)	–	–	–	Feigenbaum (1982)
	Vineyard Sound, Massachusetts	0	3.5–20.5	0.7–6.0	–	–	–	–	Feigenbaum (1982)
	Saanich Inlet, CEPEX bags	–	Mature	≥8	–	0.19	–	–	Sullivan (unpub) cited in Reeve (1980)
	Gullmarsfjorden, Sweden	6	Old Generation	1.0	–	–	–	–	Øresland (1987)
	Gullmarsfjorden, Sweden	6	New Generation	0.18 / 0.87	–	–	–	–	Øresland (1987)
Sagitta elegans and *Sagitta setosa* combined	Laboratory	15	10–22	2.04[c]	0.062	–	–	–	Kuhlmann (1977)
Sagitta enflata	Kaneohe Bay, Hawaii	24–26	4–13	7.4	–	–	0.607(2.06–0.253) (4mm–13mm) 0.704 for population	0.898(2.74–0.394) (4mm–13mm) 1.11 for population	Szyper (1976, 1978)
	Florida Current	–	12.5–20.5	2.23	0.124–0.077 (12.5mm–20.5mm)	0.264–0.144 (12.5mm–20.5mm)	–	–	Feigenbaum (1979a)
	Laboratory	21	17	10[a]	–	–	–	–	Reeve (1980)
	Virginia continental shelf	25.4	3.2–23.0	1.25	–	0.0025–0.105	–	–	Bushing and Feigenbaum (1989)
	Kaneohe bay Hawaii	–	<3.0–>10.6	8.6–18.8	–	–	0.15–2.10	–	Kimmerer (1984)
	Peru upwelling	–	9–15	0.65 / 3.2	Max=0.10 Wet wt. Basis	–	–	–	Drits (1981)
Sagitta euxina	Eastern Black Sea	–	–	2.53	–	–	–	–	Mironov (1960)
Sagitta hispida	Laboratory	24	8.5	40–50	–	–	–	–	Reeve (1964)
		24	2.5–9.5	5–60[b]	–	–	–	–	Reeve (1964)
		24	8.5	50[b]	0.64	–	–	–	Reeve (1964)
		16	6.9	10.8[c]	–	–	–	–	Reeve (1970)[a]
		21	7.0	14.1[c]	–	–	–	–	Reeve (1970)[a]
		26	6.9	23.7[c]	–	–	–	–	Reeve (1970)[a]
		24–26	larvae	–	≈1.0	–	–	–	Reeve and Walter (1972)[a]
		24–6	adult	–	≈0.10	–	–	–	Reeve and Walter (1972)[a]

Table 5.2 *(continued)*

Species	Location	Temperature (°C)	Chaetognath length (mm)	Daily ration No./day	Specific Daily Ration				Reference
					Dry wt basis	Carbon basis	Nitrogen basis	Phosphorus basis	
Sagitta nagae	Suruga Bay, Japan	–	–	0.9[c]	0.188[c]	–	–	–	Nagasawa and Marumo (1972)
Sagitta setosa	Bay of Sevastopol, Russia	20	1–10	4.8[c]	1.68–0.072[c] (larvae–adult)	–	–	–	Mironov (1960)
Sagitta tenuis	Chesapeake Bay, Virginia	21 + 25	4.5–9.5	5.36	0.358 (0.644–0.214) (4.5mm–9.5mm)	0.412 (0.809–0.220)	0.327 (0.626–0.181)	–	Canino + Grant (1985)
Sagitta setosa	Gullmarsfjorden Sweden	14	–	2.3	–	–	–	–	Øresland (1987)
Sagitta crassa	Tokyo Bay, Japan	–	–	7.1	–	–	–	–	Nagasawa and Marumo (1984)
	Laboratory	17.8–26.5	varied	8.7–10.4	0.347–0.568	–	–	–	Nagasawa (1984)

[a] maximum ingestion rate during experiments; [b] data taken from graph; [c] based on calculations using data from Feigenbaum and Maris (1984). (Updated from Feigenbaum and Maris, 1984.)

statistically significant. A few workers have found that day–night differences are reduced in winter, possibly because of reduced light intensities compared with the summer conditions (Pearre 1973; Øresland 1987). Both studies were made in fairly high latitudes. Deep sea chaetognath feeding patterns are considered in Chapter 10.

Chaetognaths that have fed late in the dark period will retain their prey well into the daytime. This is particularly true in cold waters, where digestion times are long. A careful analysis of gut contents can reveal this smearing of day–night differences (Sullivan 1980) and possibly help factor it out.

Feeding and energetics, growth, and reproduction

By comparing the energetic content of the daily ration with estimated requirements based on respiration data it is possible to make inferences about segments of chaetognath populations. Pearre (1981) found that for *S. elegans* in Bedford Basin, Nova Scotia, lack of sufficient energy in the diet of Stage III individuals in July and Stage I in December could have been responsible for the heavy mortalities noted in population surveys. Feigenbaum (1982), analysing *S. elegans* in winter at Vineyard Sound, Massachusetts, noted that small individuals consumed in excess of their requirements, which implied growth during this cold water period, while adults were feeding at the minimum rate necessary to maintain themselves.

Food contributes to the growth of immature chaetognaths and to the reproduction of egg-producing individuals (Reeve 1966, 1970a; Nagasawa 1984) while

starved animals halt their reproductive development and shrink (Reeve *et al.* 1970). Both McLaren (1969) and Sameoto (1973) found *S. elegans* reproduction unrelated to copepod biomass in studies of natural populations but concluded that timing in relation to food was more important than the food level itself. King (1979) also found chaetognath reproduction tied to the abundance of small copepods. However, Dunbar (1962) found reproduction of Arctic *S. elegans*, which have a long spawning period, was not accurately timed with food availability.

Stone (1966) found that *S. enflata* from food-rich neritic waters in the Agulhas current had more eggs per individual than those from food-poor oceanic stations. However, temperature, which can affect size at maturity (Dunbar 1962; Sameoto 1971), may have been a factor.

The significance of chaetognath feeding to planktonic food webs

The great abundance of chaetognaths in the sea was noticed by the earliest workers (Darwin 1844; Busch 1851; Grassi 1883; Chun, cited in Shipley 1901). Analysing literature records, Reeve (1970a) deduced that chaetognaths have a biomass equal to about 30 per cent of that of copepods in the world's oceans. He speculated that most of the energy converted to animal biomass by copepods was transferred to higher trophic levels via these predators.

Aside from energy considerations, chaetognaths can influence prey populations. Kimmerer (1984) examined the impact of *S. enflata* in Kaneohe Bay, Hawaii, and concluded that the chaetognath is capable of cropping a substantial part of the copepod population at certain times of the year. The chaetognath's effect on appen-

dicularians, while substantial, was less significant, and cannibalism removed only a small fraction of chaetognath production. Some other studies that have estimated the percentage of herbivore standing stock, or secondary production consumed by chaetognath populations, are listed in Feigenbaum and Maris (1984). Nagasawa and Marumo (1984b) also recently addressed this question.

Lastly, food availability may influence the distribution and zoogeography of chaetognath species. Cheney (1985), analysing chaetognath populations in continental slope and northern Sargasso Sea water masses, speculated that the reduced macrozooplankton biomass in the northern Sargasso Sea water may be responsible for limiting the slope water species from this region.

6

THE TETRODOTOXIN VENOM OF CHAETOGNATHS

Erik V Thuesen

Introduction

For many years it was suggested that chaetognaths paralysed their prey with a toxin before ingesting them. These suggestions were based on light microscopy observations (Burfield 1927; Kuhl 1938), observations of chaetognaths feeding in the laboratory (Parry 1944; Feigenbaum and Maris 1984; Nagasawa 1985c) and observations of the microstructure of chaetognath buccal morphology made by scanning electron microscopy (Bieri *et al.* 1983; Thuesen and Bieri 1987). However, because only very small quantities of toxic substance would be required, chemical evidence of a chaetognath toxin was lacking until, using a very sensitive bioassay procedure, a neurotoxin which blocked sodium channels was discovered in the heads of sagittid, spadellid, and eukrohnid chaetognaths (Thuesen *et al.* 1988a). The toxin was identified as tetrodotoxin (TTX), an extremely potent sodium channel-blocking neurotoxin which derives its name from the family of

poisonous pufferfish, Tetraodontidae, with which it is usually associated. Although several worm phyla have members which use venom to capture prey (Kem 1988), only one other kind of animal, a tropical octopod, is known to possess a TTX venom (Sheumack *et al.* 1978).

Tetrodotoxin is synthesized by several species of bacteria (Simidu *et al.* 1987; Tamplin *et al.* 1987; Yotsu *et al.* 1987), and a TTX-producing bacterium, *Vibrio alginolyticus*, has been isolated from the chaetognaths *Flaccisagitta lyra*, *Parasagitta elegans*, *Zonosagitta nagae* and *Eukrohnia hamata* (Thuesen and Kogure 1989). *V. alginolyticus* is probably responsible for production of the TTX in chaetognath venom, and chaetognaths may act as a vector in the dispersal of TTX and TTX-producing bacteria through the environment.

Identification of the toxin in chaetognaths

The presence of a chaetognath toxin was first confirmed using a simple bioassay which can detect picogram amounts of sodium channel-blocking neurotoxins (Thuesen *et al.* 1988a). A full description of the bioassay procedures can be found in Kogure *et al.* 1988a). In brief, the bioassay uses mouse neuroblastoma cell culture (ATCC No. CCL 131) and utilizes the ability of veratridine to enhance the sodium ion influx into neuroblastoma cells when Na$^+$–K$^+$-ATPase is inhibited by ouabain. TTX and other sodium channel-blocking neurotoxins are effective in blocking this enhanced influx. The cell culture bioassay incorporates this cancellation of effects, and the presence of toxin in samples

is demonstrated by cultures which remain viable in the presence of veratridine and ouabain. Using TTX as a standard, the calculation of cell death rates using phase-contrast microscopy allows the quantitative estimation of sample toxicity as described by Kogure *et al.* (1988a). With the exception of one mesopelagic species, *Solidosagitta zetesios*, analysis of body extracts were negative and toxin was only found in the heads of chaetognaths.

The primary disadvantage of the cell culture bioassay is that it is unable to distinguish between the different classes of neurotoxins which block the sodium channel. In order to identify the responsible toxin, analysis by gas chromatography–mass spectrometry (GC–MS) of extracts from the heads of the large boreal chaetognath *Parasagitta elegans* were performed following the trimethylsilylation derivative procedures of Onoue *et al.* (1984). When compared with tetrodotoxin extracted from pufferfish ovaries, indicative sharp fragment ion peaks in the mass spectrum of samples and sample retention times on the selected ion-monitored chromatograms showed the toxin in *P. elegans* to be tetrodotoxin or a tetrodotoxin analogue (Thuesen *et al.* 1988*a*).

The results of GC–MS analysis of body extracts were negative. Ion-paired reverse-phase high performance liquid chromatography (HPLC) analysis of head extracts of *P. elegans* and *E. hamata*, performed by the methods of Nagashima *et al.* (1987), resulted in peaks with elution times identical to those of both TTX and anhydro-tetrodotoxin (anhydro-TTX) (K. Hashimoto, personal communication). HPLC analysis of the body extracts failed to find either TTX or TTX analogues.

Tetrodotoxin

Tetrodotoxin (Fig. 6.1) is a low molecular weight (319.28) water insoluble compound which is soluble and highly stable in dilute acids, but destroyed by strong acids and alkaline solutions (Mosher 1986). Anhydro-TTX lacks a hydroxy group at C4, is stable in basic solutions, and is close to 100 times less potent than TTX itself (Mosher 1986). Both TTX-'plug' and TTX-'lid' models have been proposed as sodium channel-blocking mechanisms (Shimizu 1986), however, evidence for the latter mechanism appears to be stronger. The lethal dose (LD$_{50}$: amount of intraperitoneally injected toxin which results in a 50 per cent death rate of assay individuals in approximately 30 min) of TTX is considered to be 10 μg per kg body weight (Evans 1972).

For many years TTX was thought to be confined to toxic species of pufferfish, but it is now also known to occur in amphibians, gastropods, cephalopods, echinoderms, and crustaceans (Fuhrman 1986). Many animals which accumulate TTX and other sodium channel-blocking neurotoxins are resistant to their effects (Evans 1972). The electrogenic system of chaetognath

musculature is probably driven by Ca^{2+} (see Chapter 4), as it is in the musculature of many invertebrates (Savineau and Duvert, 1986; Bone *et al.* 1987*b*), and this could allow chaetognaths to have some resistance to TTX. However, Duvert and Savineau (1986) have shown high concentrations of TTX (10^{-6} g/ml) to have a partial inhibitory effect on electrically induced contractions of the trunk musculature of *Sagitta setosa*, suggesting that Na^{+} also plays a role in generating action potentials in chaetognath muscles. Mechanisms of TTX resistance are not fully understood for the other animals in which TTX is known to occur, and the susceptibility of the chaetognath nervous system to TTX is unknown.

Many *Vibrio* bacteria are capable of producing TTX and anhydro-TTX (Simidu *et al.* 1987), and bacteria are the probable source of the TTX which is found in chaetognaths. Bacteria are also the most likely source of the TTX which accumulates in marine sediments (Kogure *et al.* 1988*b*), however, it is possible that dead chaetognaths also contribute to the formation of these high TTX concentrations.

Bacterial production of tetrodotoxin in chaetognaths

TTX-producing bacteria have been isolated from four species of planktonic chaetognaths: *Flaccisagitta lyra*, *Parasagitta elegans*, *Zonosagitta nagae*, and *Eukrohnia hamata* (Thuesen and Kogure 1989). Extracts taken from batch cultures of all *Vibrio* strains isolated from chaetognath heads effectively blocked the sodium channel as determined by cell culture bioassay, and analysis by ion-paired reverse-phase HPLC of both bacterial cell extracts and culture supernatant showed that the sodium channel-blocking toxin was in fact TTX (Thuesen and Kogure 1989). The characterization of

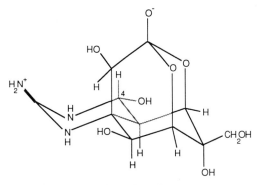

Fig. 6.1 The structure of tetrodotoxin.

bacteria isolated from *F. lyra*, *P. elegans*, *Z. nagae*, and *E. hamata* showed that all four chaetognaths possess the same TTX-producing bacterium, and every *Vibrio* strain which was isolated from chaetognath heads was identified as *Vibrio alginolyticus*. Extracellular TTX was found in the medium of *V. alginolyticus* cultures at concentrations up to 790 pg/μl within 24 hours (Thuesen and Kogure 1989). These high concentrations of extracellular TTX showed that *V. alginolyticus* is able to secrete substantial quantities of TTX in a short period of time. Although such TTX secretion rates observed in culture may indicate that *V. alginolyticus* can secrete sufficient TTX in the time required for replenishment of venom TTX following prey capture, exact TTX secretion rates and numbers of *V. alginolyticus* inhabiting chaetognaths are not yet known.

TTX in the venom of the only other animal known to use it to capture prey, the blue-ringed octopus *Hapalochlaena maculosa* (Sheumack *et al.* 1978), is also apparently produced by bacteria (Hwang *et al.* 1989). *Vibrio alginolyticus* is the only bacterium believed to be responsible for toxification of the tetraodontid fish, *Fugu vermicularis* (Noguchi *et al.* 1987) and a toxic starfish, *Astropecten polyacanthus.* (Narita *et al.* 1987). *Vibrio alginolyticus* is the most apparent source

of TTX in chaetognath venom and, although it is possible that the TTX in chaetognath venom is produced by another (as yet unidentified) micro-organism, it is remarkable that this same TTX-producing bacterium has been found in association with such an array of marine animals which possess TTX.

The exact location of *Vibrio alginolyticus* and the mechanism of accumulation of TTX in chaetognath heads is still unknown. The vestibular papillae which lie beneath the tips of the posterior teeth, the vestibular pit, and the mouth and gut are all possible locations for the bacteria (Thuesen *et al.* 1988*b*). There have been very few studies of chaetognath–bacteria associations. The work of Nagasawa and co-workers (Nagasawa and Nemoto 1984); Nagasawa *et al.* 1984, 1985*b*) focused primarily on pathogens of chaetognaths and epi-bacteria associations. Nair *et al.* (1988) found that unidentified *Vibrio* species accounted for 12.7 per cent of the bacteria associated with healthy individuals of *Aidanosagitta crassa* and 36.3 per cent of those associated with weak and moribund individuals. It is likely that *V. alginolyticus* accounted for some of these *Vibrio*. The ultrastructural characteristics of bacterial associations and the physiological mechanisms of TTX accumulation have not yet been revealed in chaetognaths or any of the other animals which possess TTX.

Prey capture and envenomation

Observations of chaetognaths paralysing copepods during laboratory feeding experiments (Parry 1944; Feigenbaum and Maris 1985; Nagasawa 1985*c*) provided the first direct evidence that chaetognaths used a venom to enhance their feeding capabilities. The impressive raptorial apparatus of a sagittid chaetognath is shown in Figure 6.2. The two sets of laterally located grasping spines are used for capturing prey and form a basket preventing the escape of prey once they are caught (Darwin 1844); the grasping spines also act to manipulate prey and force it into the gut. The posterior and anterior teeth, and the grasping spines, are capable of penetrating the epidermis of fish larvae or the crustacean exoskeleton (Thuesen and Bieri 1987). The chitinous teeth are multicuspate (Aida 1897; Furnestin 1982; Bieri *et al.* 1983) and their tips are hardened with silicon (Bone *et al.* 1983), two factors which undoubtedly aid in the puncture of prey exoskeleton and epidermis. Bone *et al.*(1983) have also shown

that the grasping spines are made from zinc-hardened chitin and have siliceous tips, but they are not cuspate in design. Neither the teeth nor the grasping spines are hollow (Bone *et al.* 1983) and therefore have no means of directly injecting venom into captured prey. However Bieri *et al.* (1983) have proposed that the papillae of the vestibular ridge may secrete a venom when they are pressed against wounds caused by the posterior teeth. These papillae apparently also have a mechanosensory ability (Bone and Pulsford 1984), which could play a role during envenomation. The secretions of the mouth, gut and vestibular pit may also be toxic, and the accumulation of TTX in some or all of these secretions could aid in the paralysing of prey as it is captured and wounded by the grasping spines and teeth. If the prey item is large or spiny, paralysis may be necessary to allow the chaetognath to orient it longitudinally before ingestion can begin. Secretions from the mouth and/or vestibular pit may also serve to

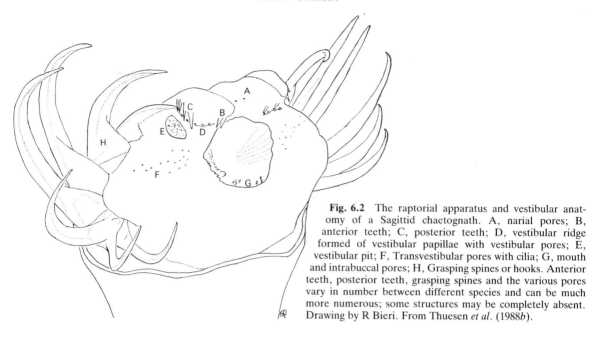

Fig. 6.2 The raptorial apparatus and vestibular anatomy of a Sagittid chaetognath. A, narial pores; B, anterior teeth; C, posterior teeth; D, vestibular ridge formed of vestibular papillae with vestibular pores; E, vestibular pit; F, Transvestibular pores with cilia; G, mouth and intrabuccal pores; H, Grasping spines or hooks. Anterior teeth, posterior teeth, grasping spines and the various pores vary in number between different species and can be much more numerous; some structures may be completely absent. Drawing by R Bieri. From Thuesen *et al.* (1988*b*).

entangle prey and act as a lubricant to prevent damage to the chaetognath as prey is forced into the gut (Parry 1944).

Nagasawa (1985*c*) made laboratory observations of *Aidanosagitta crassa* which selectively released prey copepods that were infested with bacteria and inferred that the chaetognath was able to detect the presence of bacteria via chemoreceptors in the buccal area. In addition to the vestibular ridge papillae, there are several other kinds of supposed sensory structures in the buccal area of chaetognaths (Bone and Pulsford 1984; Thuesen *et al.* 1988*b*; see Fig. 6.2) and these may also function during the selection, manipulation or envenomation of prey. However, the actual role of the narial pores, intrabuccal pores, transvestibular pores, and cilia of the transvestibular pores remains obscure at present.

Quantitative estimations of chaetognath toxicities were carried out using the cell culture bioassay described previously (Table 6.1). Toxin was not found in the bodies of chaetognaths, except in *Solidosagitta zetesios*, which is also the chaetognath with the highest toxicity. Toxin may also be present in the bodies and/or guts of the other chaetognaths, but at levels beyond the limit of detection. Nevertheless, head toxicities of *S. zetesios* were of an order higher than those in the body, indicating that the location of TTX accumulation is primarily in this region. This further supports the

idea of the toxin functioning as a venom used to capture prey. The TTX in chaetognaths could also serve as a defence against predation, but this has not been investigated.

LD_{50} estimates of TTX for chaetognath prey indicate that chaetognaths possess ample quantities of toxin to induce paralysis in their prey. Assuming an LD_{50} of 10 μg TTX per kg prey body weight, 10 pg TTX is required to paralyse a copepod with a wet weight of 1 mg. Analyses of chaetognath prey size (Pearre 1980*a*) and the wet weights of copepods in relation to body size parameters (Pearre 1980*b*) have shown that several common species of chaetognaths regularly eat copepods of less than 1 mg, although they may be capable of consuming larger ones. There is evidence that copepods are susceptible to the effects of sodium channel-blocking neurotoxins—paralysis results after feeding on red tide dinoflagellates (Ives 1985), which are known to contain various sodium channel-blocking neurotoxins. The toxicity data in Table 6.1 suggests that chaetognaths have substantially more than enough toxin to be effective predators on copepods, since larger doses of toxin result in faster effects. Chaetognaths should also be able to paralyse numerous other prey, but analyses of feeding rates have shown that they rarely appear in plankton samples with more than one or two prey items in the gut (Feigenbaum and Maris 1984; See Chapter 5).

Table 6.1   Estimated tetrodotoxin toxicities of some common chaetognaths[1]

Taxa	Body length (mm)	Toxicity (pg/individual)
Eukrohnidae		
Eukrohnia hamata	25	140
Sagittidae		
Aidanosagitta crassa	15	30
Flaccisagitta enflata	18	75
F. hexaptera	44	245
F. scrippsae	60	290
Parasagitta elegans	38	320
Solidosagitta zetesios (head)	47	5200
Solidosagitta zetesios (body)	47	764
Zonosagitta nagae	28	262
Spadellidae		
Spadella angulata	6	60

[1]Estimated toxicity data from Thuesen *et al.* (1988*a*) except those for *Flaccisagitta hexaptera*, *Zonosagitta nagae*, and *Solidosagitta zetesios*, which were estimated by cell culture bioassay (Kogure *et al.* 1988*a*) following procedures described in Thuesen *et al.* (1988*a*). Toxicity estimates are from pooled samples and do not account for individual variation.

In general, toxicities tend to increase with body size (Table 6.1). The most toxic chaetognath, *Solidosagitta zetesios*, is large and very muscular, and it may use its large dose of TTX to capture large prey or chaetognaths that may be partially resistant to TTX. Many species of chaetognaths are known to be cannibalistic feeders (Pearre 1982), and whether their TTX venom is of practical use against other arrow worms remains to be seen. Laboratory observations and physiological investigations have indicated that some species of chaetognaths may be 'ambush predators' (Feigenbaum and Maris 1984; Bone *et al.* 1987*a*), and the use of a TTX venom would greatly enhance the success rate of this type of predator strategy. If predator–prey encounters are few, paralysis of prey should ensure successful prey capture.

Concluding remarks

Although chaetognaths present no direct threat to human health, the ubiquity of chaetognaths suggests that they may play a role in the dispersal of TTX and TTX-producing bacteria through the marine environment, and their role in this respect deserves further attention. Investigations of bacterial interactions with chaetognaths, as mentioned in this Chapter and by Nagasawa (Chapter 8), have not yet progressed beyond preliminary stages. Studies by microbiologists on the TTX-producing capabilities of bacteria and the role of TTX in bacteria physiology should provide some direction for investigations of TTX–bacteria–animal relationships. The symbiotic nature of such relationships has yet to be shown.

Much work remains to be done on the interaction of tetrodotoxin and chaetognath physiology. Studies on the unique nervous systems and neuromuscular systems (Chapters 3 and 4) of chaetognaths may eventually allow us to understand the adaptions which let chaetognaths and other animals accumulate relatively large quantities of sodium channel-blocking neurotoxins.

The mechanics of prey envenomation remain obscure because of the scarcity of visual observations of chaetognaths capturing prey. Perhaps work using high speed microcinematography could help to rectify this (as it has for the study of copepod feeding mechanics). It is not yet known how many of the approximately 100 species of arrow worms use a TTX venom during prey capture. The phylogenic diversity of the known venomous species suggests that possession of TTX may be very common throughout the phylum Chaetognatha and indeed it would be very interesting to discover if rare bathybenthic toothless chaetognaths, such as *Krohnittella tokiokai*, (Bieri, 1974a) also have tetrodotoxin venoms.

Acknowledgements

I am very grateful to the late Dr T. Nemoto, to Drs K. Hashimoto, K. Kogure, S. Nagasawa, M. Terazaki, and all of my colleagues at the Ocean Research Institute, University of Tokyo for their encouragement and support. The Ministry of Education, Science and Culture, Japan (*Monbushô*) financially supported much of my research on tetrodotoxin and chaetognaths. I am deeply indebted to the late Professor Robert Bieri, Antioch College, for introducing me to the Chaetognatha and providing a continuous source of intellectual stimulation.

7

GROWTH AND REPRODUCTION
Sifford Pearre, Jr.

Introduction

Interest in the anatomy and physiology of growth and reproduction of the Chaetognatha dates from the middle of the last century. Much of the early investigation addressed questions about their phylogenetic relationships, central to which was the discovery that the mouth was not derived from the original blastopore opening: this placed the chaetognaths among the evolutionarily 'advanced' Deuterostoma. However, Deuterostomes may have been polyphyletic (Salvini-Plawen 1988), so this discovery may be less significant than formerly believed. Recent decades have seen a rise in research on growth and ecological production, spurred by advances in culture techniques and ecological theory. General treatises on the phylum include Burfield (1927), Kuhl (1938), Hyman (1959), de Beauchamp (1960), Alvariño (1965), Ghirardelli (1968), and Boltovskoy (1981b). There are also several recent reviews devoted specifically to chaetognath reproduction, including Ghirardelli (1959a, b), Reeve and Cosper (1975), Alvariño (1983a, b), and Strathmann and Shinn (1987). In view of the richness and ready availability of this material, and the limitations of space, this review concentrates on growth and production and their relationships with the environment; which have not been so extensively surveyed.

Nomenclatural notes

Early identifications of *Sagitta bipunctata* in north European waters (Burfield 1927) probably refer to *S. setosa* or *S. elegans* (Russell 1931). Alvariño (1983a) believes that *S. bipunctata* identified by workers in southern European waters should be referred to *S. friderici*; to avoid confusion I will use the original authors' designations but the reader is warned to be cautious. '*S. bipunctata*' from the US west coast (Michael 1911) should probably be *S. euneritica* (Alvariño 1965) and will be so designated.

S. euxina may be a synonym of *S. setosa* (Furnestin 1961); both names are used here.

S. gazellae might be synonymous with *S. lyra* in high latitudes (Tchindonova 1955) and/or with *S. scrippsae* (Tokioka 1974b; Casanova 1977 (in Michel 1984)). I will use all three names to avoid confusion in this review.

Many authors refer to *S. inflata* (e.g. Stevens 1910; Ghirardelli 1968; Reeve and Cosper 1975); the original and nomenclaturally correct spelling is *S. enflata* (Grassi 1881) and this is used here.

'*Spadella draco*' of some early works (e.g. Hertwig 1880b, c; Michael 1911; Sanzo 1937) is correctly *Pterosagitta draco*, a planktonic species (*Spadella* is a benthic genus).

Eukrohnia subtilis as used by Michael (1911) is amended to *Krohnitta subtilis* (Michel 1984).

Anatomy of reproduction and fertilization

Chaetognaths are hermaphrodite, and their reproductive anatomy is described in Chapter 2. There is no known asexual reproduction. Fertilization is internal (see p. 62), but the exact nature of the process once the sperm are inside the recipient chaetognath is still not well understood because knowledge of the anatomy of the ovaries is incomplete (Strathmann and Shinn 1987). Spermatocytes are produced in the caudal region behind the transverse caudal septum. This region is divided by a longitudinal septum, which begins at the caudal septum and runs to the tail. In *Sagitta* species there are also incomplete lateral subsepta, allowing communication at both ends (Burfield 1927; Reeve and Cosper 1975; Alvariño 1983a). Immature spermatocytes are in constant motion in the region of the testes, moving aft along the median longitudinal septum and forward again between the body wall and the subsepta (Burfield 1927; Jägersten 1940, de Beauchamp 1960 (*S. setosâ*); Alvariño 1983*a* (*Sagitta* species)). This motion is probably caused by the action of cilia on the medial septum (Darwin 1844; Burfield 1927; Ghirardelli 1968; Reeve and Cosper 1975; Alvariño 1983*a*).

The sperm is filiform (Bolles Lee 1887; Stevens 1903; Tuzet 1931; Jägersten 1940 (*S. bipunctata*); Nagasawa 1987*b*; (*S. crassa*); Bolles Lee 1887 (*S. minima*); de Beauchamp 1960 (*S. setosa*); Hertwig 1880*b,c*; Tuzet 1931; van Deurs 1972 (*Spadella cephaloptera*); Goto and Yoshida 1985 (*S. schizoptera*)). The ultrastructure of chaetognath sperm differs from that of filiform sperm of other groups, for example, arthropods and gastropods (van Deurs 1972; Alvariño 1983*a*).

Mature spermatozoa exit from the testes via a ciliated vas deferens (Hertwig 1880*b, c*) into seminal vesicles which protrude from the body wall in the caudal region just ahead of the caudal fin (tail). These structures, especially when packed with mature sperm, have distinctive sizes, shapes and locations which have been used for species identification (mature stages only). Alvariño (1983*b*) provides a useful compendium of seminal vesicle morphology for pelagic species.

Once in contact with the body surface, the sperm moves rapidly towards the opening to the ovary (Stevens 1910; Bordás 1920; Jägersten 1940). Ghirardelli (1968, *S. cephaloptera*, *Sagitta enflata*, *S. bipunctata*), Reeve and Walter (1972*a*, *S. hispida*), Nagasawa (1985*b*, *S. crassa*), and Goto and Yoshida (1985, *Spadella schizoptera*) reported that both the ability of the sperm cluster to adhere and the ability of the sperm to find the seminal receptacles are affected by the position of the sperm cluster on the receptor animal. The precision necessary for sperm cluster placement has been used as an argument against self-fertilization. Ghirardelli suggested that a chemotaxis was involved in the sperm movement.

As far as is known, all chaetognaths are protandrous hermaphrodites—the testes mature before the ovaries (Hyman 1959; Dunbar 1962; Ghirardelli 1968; Reeve 1970*a*; Strathmann and Shinn 1987). There has been considerable disagreement over whether individuals self-fertilize or must cross-fertilize. Thomson (1947) noted differences in the degree of protandry among pelagic species. An extreme protandry would make self-fertilization impossible, but Reeve (1970*a*) felt that in general the difference in timing is slight. Self-fertilization has been experimentally induced in *Sagitta setosa* (Dallot 1968), *S. hispida* (Reeve 1970*a*, Reeve and Walter 1972*a*), *S. elegans* (Pearre unpublished

observation), *S. crassa* (Nagasawa 1987*b*), and *Spadella cephaloptera* (Ghirardelli 1968), so it is possible under laboratory conditions. A number of authors have concluded that self-fertilization is usual or necessary (Stevens 1910; Bordás 1920) *Sagitta bipunctata*); Jägersten 1940 (*S. setosa*); Hyman 1959 (*Sagitta* spp.). Others feel that in nature, cross-fertilization is obligatory either because of physical difficulties in bringing sperm to the seminal receptacle (above) or because of evolutionary arguments (Grassi 1883; Conant 1896; van Oye 1931; Vasiljev 1925; John 1933; Ghirardelli 1968 (*Spadella cephaloptera*); Alvariño 1983*b* (all species). Conant (1896) and Reeve and Walter (1972*a*) argued against self-fertilization in nature because, although the first batch of eggs laid by isolated *Sagitta hispida* developed normally, subsequent batches proved infertile, as might be expected if the animal had received an initial sperm supply before isolation and had subsequently run out of sperm. However, Dallot (1968) reported that 50 per cent of the eggs laid by isolated *S. setosa* were fertile, and Nagasawa (1984) reported that *S. crassa* which had been isolated since before maturity laid fertile eggs.

Although mating has been often observed in *Spadella*, it has been seen very seldom in pelagic chaetognaths (van Oye 1931 (*Sagitta* sp. *S. elegans* or *S. setosa*); Murakami 1959 (*S. crassa*); Reeve and Walter 1972*a* (*S. hispida*)). Casanova's (1985*c*) report of internal connections between testes and ovaries in *Heterokrohnia murina* and *H. davidii* appears to indicate a system designed for self-fertilization, assuming a low degree of protandry. For most species, however, the consensus is that cross-fertilization is more usual in nature (Alvariño 1965, 1983*b* (all species); Kotori 1979 (*S. elegans*); Ghirardelli 1968; Reeve and Walter 1972*a*; Reeve and Cosper 1975 (*S. hispida*); Nagasawa 1985*b* (*S. crassa*); Goto and Yoshida 1985 (*Spadella schizoptera*)).

Upon arrival at the seminal receptacle, the sperm enters and reaches the vicinity of the eggs either by a separate duct or via the combined 'ovispermaduct'. Reeve and Walter (1972*a*) describe the stream of sperm dividing so that some enter each gonopore (*Sagitta hispida*). There the sperm pass through a stem-like structure (accessory fertilization cells) to fertilize the egg (Reeve and Lester 1974). Stevens (1910), Ghirardelli (1968), and Reeve and Cosper (1975) have reported that the egg is fertilized while still in the ovary; Alvariño (1983*b*) feels that fertilization occurs after extrusion of both eggs and sperm into the vaginal cavity.

Egg laying

The question of how the fertilized eggs (zygotes) are expelled is also unresolved. Stevens (1910), *S. bipunctata*); John (1933, *Spadella cephaloptera*); and Ghirardelli (1959a, 1968, *Sagitta bipunctata* and *S. enflata*) have suggested that the eggs move actively by ciliary or amoeboid action. Conant (1896) and Reeve and Lester (1974) (*S. hispida*) and Dallot (1968), (*S. setosa*) feel that muscular effort by the adult *S. hispida* forces them out.

It is not clear what route the zygotes follow to the outside, nor if this is the same in all species. In *Sagitta* species zygotes travel in either the ovispermaduct (Reeve and Lester 1974 *S. hispida*) or a temporary oviduct (Stevens 1910; Ghirardelli 1968 *S. bipunctata* and *S. enflata*). In *Spadella* the eggs appear to move through a temporary canal within the vagina (Ghirardelli, 1968).

Eggs of the benthic *Spadella* exit singly and are attached, in clusters of 4–12, by an 'elastic peduncle' to the grasses or seaweeds of the substratum (Ghirardelli 1968). Eggs of *Sagitta hispida* are extruded as a double string from each gonopore and are also attached to the substratum, although in larger clumps and by an adhesive coating (Conant 1896; Reeve and Lester 1974; Reeve and Cosper 1975). As far as is known, all other *Sagitta* species shed eggs freely into the water column. Nagasawa (1984) reported that eggs of *S. crassa* are shed singly into the water in the laboratory. It is not clear how many species have eggs which remain pelagic. Doncaster (1902) and Reeve and Cosper (1975) reported that the eggs of *S. enflata* sink to the bottom of a culture vessel, but the eggs of *S. elegans*, *S. setosa*, *S. bipunctata*, and *S. euxina* (and probably most other species) are approximately neutrally buoyant and remain in the water column (Hyman 1959; de Beauchamp 1960). Alvariño (1983a) provides a useful list, including egg sizes of pelagic species.

In *Pterosagitta draco*, 200 to 300 eggs are clumped in pelagic gelatinous masses about 6 mm across. They hatch within this mass and are liberated from it as larvae in about 4 days (Sanzo 1937).

Eukrohnia appears to retain zygotes in brood pouches or sacs (marsupia) adhering to the posterior lateral fins. Alvariño (1968) reviewed this and suggested that similar structures are found in the related genus *Krohnitta* and in the deep-sea benthic genus *Bathyspadella*.

Eggs are predominantly shed during the night (Reeve and Cosper 1975). Conant (1896) reported that *S. hispida* and *S. hexaptera* shed eggs at sunrise, and Wimpenny (1938) and Dallot (1968) reported the same for *S. setosa*. Conant (1896) and Stevens (1910) reported that *S. bipunctata* shed eggs at sunset, but Doncaster (1902) found this species to lay eggs 'in the early morning'. Stevens (1910) felt that *S. elegans* probably had no preferred egg-laying period, but Kotori (1975a, b) found that his Pacific specimens shed them at 20.00 hours. Nagasawa and Marumo (1978) reported that *S. nagae* sheds its eggs before dawn: this appeared to be related to the timing of copulation, which occurred at night and which Nagasawa and Marumo (1984b) and Nagasawa (1985b) also reported in *S. crassa*. Ghirardelli (1968) reported that *S. enflata* shed eggs between 23.00 hours and dawn but that, although night was favoured, *Spadella cephaloptera* (benthic) could lay eggs at any time. However, Vasiljev (1925) felt that *Spadella* spawned only at night.

Chaetognatha have generally been considered to be semelparous—to spawn once and then die (McLaren 1969), and to cease somatic growth upon reaching maturity (see p. 66). However, this may not be true for all species. Reeve (1970a) documented continued growth (in the laboratory) of *Sagitta hispida* during the egg-laying period, and Nagasawa (1984) confirmed this in *S. crassa*. Michael (1919), Thomson (1947), Ghirardelli (1951), Furnestin (1953), Owre (1960), Boltovskoy (1975), and Koszteyn (1983) have reported (from field data) that *S. enflata* appears to go through two or more complete spawning cycles at different body sizes (reviewed by Furnestin 1957). Due to the difficulty of identification of particular populations and the likelihood of the existence of a number of subspecies (Furnestin 1957), verification of multiple cycles from field data alone is problematical. Furnestin (1957) speculated that *S. lyra* might also reproduce more than once, which Owre (1960) supported for this species and seven others. Furnestin (1961) felt that *S. euxina* in the Black Sea was actually a second maturity cycle of *S. setosa* approximately 22 mm as opposed to approximately 16 mm. Nair (1974) has reported two maturity cycles, at approximately 8 mm and 12 mm, in *S. bedoti* but Alvariño (1965) cautions that many differences in size at maturity may be due to advection and seasonal variations.

Spawning is unlikely to be a single deposition of eggs in any species. Conant (1896) remarked that an isolated

S. hispida had laid eggs (24 to 68 per day) every other day (except for the first two, which were consecutive) over a 6-day period. Reeve (1970a) reported up to nine separate batches of eggs, laid every second day.

Murakami (1959) reported that *S. crassa* laid eggs 'more than twice at intervals of several days'. Nagasawa (1984) reported egg production on 30 consecutive days. Dallot (1968) found that *S. setosa* laid up to six successive batches of eggs, separated by intervals of about 24 hours. Some of these authors also documented concurrent changes in ovary length and apparent maturity, and in fullness of the seminal vesicles. None of the animals lived past this period of egg-laying, so it is not yet possible to know whether they can go into a more protracted phase of regressed sexual maturity and continued somatic growth (see p. 000).

Wimpenny (1937) speculated that *S. setosa* spawned over a protracted season. Jakobsen (1971), King (1979) and Conway and Williams (1986) stated that *S. elegans* must spawn several times over a period of months. It is perhaps a semantic problem as to whether such behaviour constitutes one, or more than one, 'brood' (Reeve and Cosper 1975), although it could produce more than one cohort. However, these protracted and cyclical changes obviously make it far more difficult than was previously assumed to distinguish individual ages, spawning times and stocks.

Embryology

Early embryonic development has always been of special interest to workers hoping to unravel the mysteries of the systematic position and affinities of chaetognaths (see p. 61). Murakami (1959) published microphotographs of early egg divisions in *S. crassa*; Reeve and Cosper (1975) did the same for *S. hispida*. Kuhl and Kuhl (1965) present single frames from micro-time-lapse movies of the development of *S. setosa* from single-cell to completed gastrulation and the closure of the protostoma. Cleavage is total (holoblastic), equal, radial and indeterminate (Hyman 1959). However, the germ-cell line is determined at a very early stage (Stevens 1910; Hyman 1959) but by different processes in *Sagitta* and *Spadella* (Ghirardelli 1968; Strathmann and Shinn 1987). A small blastocoel is present in the blastula. However, this blastocoel is obliterated in gastrulation. A central and two lateral cavities are formed by growth of two folds in the archenteron, thus the embryonic development differs from the typical deuterostone pattern (Hyman 1959). The blastopore closes off but later becomes the site of the anus, and the mouth is formed by a stomodeal invagination which fuses with the central (gut) cavity, and the lateral cavities disappear (see also p. 16).

The time to hatching of the eggs is generally 2–3 days, depending on species and environmental conditions (see Table 7.1).

Growth and Development

Maturity stage classification

Most studies of chaetognath ecology categorize animals into some maturity classification scheme for convenience, and it is important to know what criteria have been used in a given study.

Development classification is generally based on various combinations of ovary and/or testis development. Some schemes depend on staining techniques (Russell 1932a), reducing their desirability for field use and also perhaps damaging the specimens (Sands 1980). Ghirardelli (1959b), Alvariño (1965), Reeve and Cosper (1975), and Boltovskoy (1981b) have reviewed most of the proposed systems. Perhaps the most popular recent system has been a simplified version of Russell's (1932a) scheme, omitting staining and based solely on ovary development, as suggested by Thomson (1947). However, Reeve (1970a) and Reeve and Cosper (1975) have suggested that the periodic shedding of eggs in a protracted spawning session makes ovary development an unreliable index of maturity, and that the condition of the seminal vesicles may be a better indicator. On the other hand, Nagasawa (1984) reported that *S. crassa* had bouts of spawning lasting

Table 7.1   Larval growth in Chaetognatha

Species	Egg size (mm)	Time to hatch (h)	Size at hatch (mm)	Size at first feeding (mm)	Age at first feeding (days)	Reference
Sagitta bipunctata	0.2	*	0.51	–	8–9	Doncaster (1902)
S. crassa	0.35	27	0.8	–	–	Murakami (1959)
S. enflata	0.2	*	1	–	8–9	Doncaster (1902)
S. elegans	0.31–0.34	–	–	–	–	Zo (1973)
	0.3	67.2	1.23	–	–	Kotori (1975*a,b*)
	0.33	–	1.05	1.7	10	Kuhlmann (1976)
	–	–	–	1.28	–	Pearre (1980)
S. friderici	–	–	0.5	2.2–2.5	–	Halim and Guerguess (1973)
S. hispida	–	36	–	–	4–5	Conant (1896)
	–	<24	0.9	1.3	3–5	Reeve and Walter (1972*b*)
	–	–	0.9	–	2–3	Reeve and Cosper (1975)
S. nagae	0.16–0.28	–	0.506	–	–	Nagasawa and Marumo (1978)
S. setosa	0.16–0.2	19	0.8	–	–	Dallot (1968)
Pterosagitta draco	0.36–0.4	–	1.4	2.16	11	Sanzo (1937)
Spadella cephaloptera	0.3	48	1.5	–	7	John (1933)

* increases as temperature decreases.

several days, with a longer term periodicity of 7 to 10 days. The fullness of the seminal vesicles and ovaries varied between these sets of spawning bouts.

Reeve (1970*a*), Reeve and Cosper (1975) and Kotori (1976) have also proposed classification schemes for early (larval and juvenile) development of chaetognaths.

The 'larval' period

Newly hatched chaetognaths are 0.8–1.5 mm long, depending on species and perhaps environmental influences (Table 7.1). Although they do not undergo a metamorphosis, the early developmental changes are extensive enough to have warranted many authors calling them 'larvae'. (See Chapter 2, for a discussion of the status of chaetognath 'larvae'). Chaetognath larvae have been figured by Hertwig 1880*b* (*Sagitta* sp.); Doncaster 1902 (*S. bipunctata*, *S. enflata*); Burfield 1927 (*S. setosa*); John 1933 (*Spadella cephaloptera*); Sanzo 1937 (*Pterosagitta draco*); Murakami 1959 (*Sag-*

itta crassa); Kotori 1975*a, b*, 1976 (*S. elegans*); Kuhlmann 1976 (*S. elegans*); Reeve and Cosper 1975 (*S. hispida*); and Nagasawa and Marumo 1978 (*S. nagae*). Newly hatched larvae lack anterior fins, and caudal septum, and anus, and (except *S. hispida*, Reeve and Cosper (1975), and *Sagitta nagae*, Nagasawa and Marumo (1978)) also lack eyes and head armature (chaetae). The larvae of the benthic *Spadella* adhere to the surface on which they were hatched (John 1933), while larval *Sagitta* are planktonic, generally remaining near the water surface. *S. hispida*, being more developed at hatching, begins to feed much earlier than *S. bipunctata*, *S. enflata*, or *S. elegans* (Table 7.1).

Post-larval growth

Growth and development after attainment of juvenile form are straightforward. Meek (1928) noted that most body sections increase in direct proportion to overall length, except the section behind the caudal septum which becomes relatively shorter as the animal grows.

At the time of formation of the septum it can be nearly half of the total length, and it remains a very large proportion in *Spadella* and in *Pterosagitta*.

The other notable exception to proportional growth is the size of the ovaries. Ovaries develop from the germinal primordia which have been pushed to the rear of the body cavity by the progression of the folds of the archenteron (primitive cavity). Alvariño (1983a) provides an extensive list of lengths of the mature ovaries as a proportion of total mature length. These ratios have been used as species identification characters (Rózanska 1971), but several difficulties arise when they are used in this way:

1. Although body growth is generally thought to slow or to stop with the onset of maturity (Alvariño 1965; Sameoto 1971), this may not always be so see p. 63.

2. Mature length is influenced by environmental conditions during development, especially by food abundance and temperature (see p. 68), and larger specimens usually have *proportionally*, as well as absolutely, larger ovaries.

3. During the spawning period, the ovaries may periodically contract and then enlarge as groups of ripened eggs are extruded (Reeve 1970a).

Within a species, mature ovary length increases as an exponential function of the body length at which the chaetognath matures (Russell 1932a, b; McLaren 1963, 1966; Sameoto 1971) (an exception is one report on *S. friderici* by Halim and Guerguess 1973). McLaren (1963) found that egg number in *S. elegans* increases linearly with ovary size.

The dual process of relating egg number to ovary length, and the latter to mature body length still leaves unresolved the problem of periodic egg shedding. Several authors have addressed this problem by relating direct counts of all eggs to body length; regressions based on these data are presented in Table 7.2.

Allometrics of growth

Chaetognaths tend to grow allometrically, i.e. Their weight increases proportionally to the length raised to an exponent (not equal to three). Table 7.3 lists estimations of the allometric regression parameters for chaetognath growth, supplied by various authors or derived from their data. Occasional data on other types of weight expressions, such as carbon weight, nitrogen weight, ash-free dry weight (= 'organic weight') are also available: the reader can obtain many of these from references in this table and from Chapter 9.

Table 7.2   Relationships of total egg number to mature chaetognath length in various species, as power curves*:
$N_E = a \, L_{TM}^{b}$

Sagitta species	Area	a	b	r^2	Reference
S. crassa 'C' type	Japan Inland Sea	0.0558	2.1	0.077	Murakami (1959)
S. enflata	Japan Inland Sea	0.0111	2.94	0.91	Murakami (1959)
	South Africa, inshore				Stone (1966)
	summer	0.0483	2.5	0.95	
	fall	0.0536	2.47	0.92	
	winter	0.0722	2.31	0.85	
	all	0.14	2.11	0.86	
	offshore				
	summer	0.244	1.77	0.98	
	fall	0.464	1.37	0.74	
	winter	2.11	0.7	0.43	
	all	0.293	1.6	0.85	
S. elegans	Eastern and Arctic Canada	0.115	2.46	–	McLaren (1966)†
S. minima	Japan Inland Sea	0.18	1.91	0.94	Murakami (1959)

* N_E = number of eggs; L_{TM} = total length of mature chaetognath (mm)
† estimated line of maximum egg number for length.

Table 7.3 Allometrics of chaetognath growth: wet weight (W_w, μg) or dry weight (W_D, μg) as a function of total length (L_T, mm). Of form $W = a\,L_T^b$. Regressions of ordinary (predictive) form unless otherwise noted

Species	a	b	(Form)*	Area	Reference
Dry weight					
Sagitta crassa	0.106	3.24		Japan Inland Sea	Uye (1982)
	0.197	3.01		Suruga Bay, Japan	Nagawasa (1984)
Sagitta enflata	0.314	2.84		Kaneohe Bay, Hawaii	Peterson (1975)
	0.253	2.92		Kaneohe Bay, Hawaii	Szyper, (1976)
Sagitta elegans	0.970	2.36		St Margaret's Bay, Canada	Sameoto (1971) (G.M.)
	0.855	2.40	G.M.		
	0.114	3.08		North Pacific	Kotori (1976)
	0.350	2.73	N.-R.	Korsfjord, Norway	Matthews and Hestad (1977)
	0.064	3.30			
	0.239	2.69	G.M.	Barents Sea	Bogorov (1939)
	0.111	3.20		Celtic Sea	Conway and Williams (1986)
Sagitta gazellae	0.110	3.002		Prydz Bay, Antarctica	Ikeda and Kirkwood (1989)
Sagitta hispida	0.298	3.21		Florida, USA	Reeve and Baker (1975)
Sagitta tenuis	0.803	3.61		Chesapeake Bay, USA	Canino (1981)
Eukrohnia hamata	0.320	3.00	N.-R.	Korsfjord, w Norway	Matthews and Hestad (1977)
Eukrohnia bathypelagica					
stages 0–II:	0.72	3.20	N.-R.	Korsfjord, w Norway	Matthews and Hestad (1977)
stages III–IV:	5.700	2.20	N.-R.	Korsfjord, Norway	
'White Sea chaetognaths'	0.68	2.32		White Sea	Kosobokova (1980)
'Arctic Basin chaetognaths'	0.0039	3.95		Arctic basin	
Wet weight (W_w)					
Sagitta elegans	10.000	2.31		St Margaret's Bay, Canada	Sameoto (1971)
	10.700	2.37	G.M.		
	3.34	2.63	G.M.	Barents Sea	Bogorov (1939)
	1.81	2.79		Okhotsk/Bering Seas	Bogorov (1957)
	1.44	2.87	G.M.		
Sagitta tenuis	0.507	3.01	G.M.	Chesapeake Bay, USA	Canino (1981)
Sagitta gazellae	2.29	2.974		Prydz Bay, Antarctica	Ikeda and Kirkwood (1989)
Eukrohnia hamata	0.283	3.41	G.M.	Barents Sea	Bogorov (1939)
	2.99	2.64		Okhotsk/Bering Seas	Bogorov (1957)
	2.57	2.69	G.M.		
'large *Sagitta*'	0.42	3.33		Adriatic Sea	Shmeleva (1965)
'small *Sagitta*'	2.64	2.39		Adriatic Sea	
'*Sagitta*'	0.270	3.49		Okhotsk/Bering Seas	Lubny-Gertzyk (1953)
	0.323	3.50	G.M.		
'chaetognaths'	2.89	2.70		East Pacific	Miller (1966)

* N.-R. = Newton-Raphson form (Matthews and Hestad 1977); G.M. = geometric mean regression (Ricker 1975)

Environmental influences on growth and development

Temperature effects

Egg development rates increase with temperature (Doncaster 1902; Murakami 1959). Rates of somatic growth ('growth rates') of Chaetognatha and development rates (the rates of sexual maturation, taken here to be the inverse of generation length) are usually determined from field data because of the difficulty of maintaining populations in the laboratory (except the work of Reeve and co-workers on *S. hispida*). Sameoto (1971, 1973) and Reeve and Walter (1972*b*) have shown that development rates increase with temperature in *S. elegans* and *S. hispida*, respectively.

Growth rates also increase with temperature (Sameoto 1971, 1973; Sands 1980; Conway and Williams 1986 (*S. elegans*); Reeve and Walter 1972*b*; Reeve and Baker 1975 (*S. hispida*); Sands 1980 (*Eukrohnia hamata*)).

Size at maturity also changes as a function of temperature; presumably because of a differential response of growth and development rates to temperature. In general, chaetognaths mature at larger sizes in lower temperatures (Russell 1932*a*, *b*; Hirota 1959; Murakami 1959; McLaren 1963, 1966; Sameoto 1971, 1973; Reeve and Walter 1972*b*; Zo 1973; Reeve and Baker 1975; Pearre 1976*a*; Koszteyn 1983).

Because of its relationship to mature length (above), fecundity is also an inverse function of temperature. Stone (1966) examined the fecundity of *S. enflata* in three seasons and between areas of higher (offshore) and lower (inshore) temperatures (See Table 7.2). The inshore populations were larger at maturity and also more fecund at mature length; Stone ascribed part of the difference to the temperature difference and part to superior food conditions inshore (see p. 69). Koszteyn (1983) also reported a fecundity difference between warm and cool water populations of this species. Sameoto (1971) and Tiselius and Peterson (1986) reported that egg numbers were inversely related to developmental temperature in *S. elegans*.

Other morphological changes

Sagitta crassa undergoes remarkable changes in both size and morphology when it develops in different areas and seasons. The large form (*S. crassa* f. *crassa*) is limited to winter, and the transformations between this, the small summer form (*S. crassa* f. *naikaiensis*), and a large number of intermediate forms, are probably mainly temperature-determined (Kado 1954; Kado and Hirota 1957; Hirota 1959; Murakami 1959, 1966). Tokioka (1974*a*) reported yet another ecophenotype, *S. crassa* f. *tumida* (as *Aidanosagitta*) from a different area. Based on the morphological plasticity of this species, Tokioka (1974*a*) speculated that some other species pairs or groups may be really seasonal or latitudinal morphs of each other (see p. 131 and Tchindonova 1955). It must be borne in mind that the *S. crassa* complex probably only came to light because it is a culturable, coastal species in a highly variable environment with many semi-isolated local populations: such relationships would be difficult to detect in the open sea.

Effects of food abundance and size

Effects of food are much less well-documented than temperature effects. Temperature is easy to measure and changes slowly and predictably over large distances or seasons: we can be relatively certain of the temperature to which a given population has been exposed. For a rigorous examination of the influences of food, one should know the acceptable prey sizes and species, and the availability of prey for each developmental stage of each species. Although this is a counsel of perfection, nonetheless food supply is very important and should be seriously addressed in both laboratory and field studies.

Wimpenny (1937) noted that the growth rate of *S. setosa* in the North Sea apparently decreased in midsummer despite high temperatures; he attributed this to decreased food supply. Reeve (1970*b*) stated that both the concentration and composition of food affected the growth rate of *S. hispida* in culture. Rózanska (1971) thought that food supply was important to growth rate in *S. elegans baltica*. Alvariño (1983*c*) felt that poor food conditions may have retarded growth of *S. scrippsae* in the California current.

McLaren (1963) suggested that food supply may have affected generation length of *S. elegans* off Plymouth, UK, as determined by Russell (1932*a*).

Russell (1932*a*) speculated that food supply might also affect the size of maturity of *S. elegans* off Plymouth. Rózanska (1971) felt that food was important in determining the final size of *S. elegans baltica*, which lives in deep basins of relatively constant temperature. Rao and Kelly (1962) found size variations in *S. enflata* in Lawson's Bay, India, which were unrelated to temperature variations but strongly correlated with seasonal changes in copepod abundance.

Dallot (1968) reported that nutritional conditions affected the rate of egg production and release from *S. setosa* in culture. Stone (1966) concluded that higher food supply contributed more to the increased fecundity of *S. enflata* off South Africa than temperature differences (see Table 7.2).

Chaetognaths have relatively low lipid storage (Reeve *et al.* (1970) and instead appear to store proteins as an energy source. When starved they may decrease in length and/or absorb gonadal tissue (Reeve 1970*b*). Reeve (1970*a*), Pearre (1976*b*), King (1979), and Nagasawa (1984) have also noted that they may shrink after egg discharge.

As chaetognaths swallow prey whole (see Chapter 5), the maximum size of their prey is limited by their mouth size. Insufficient small prey may inhibit successful chaetognath recruitment (Kotori 1979; Sullivan 1980; Pearre 1981; Tande 1983; Conway and Williams 1986). The preferred food particle size increases as the chaetognath grows (Rakusa-Suszczewski 1969; Reeve and Walter 1972*b*; Pearre 1980*a*). Pearre (1982) suggested that, because of energetic considerations, scarcity of large prey could limit chaetognath final size even in a food-rich environment.

Effects of salinity

Murakami (1959) reported that salinity (as chlorinity) affected the time to hatching of *S. crassa* (his Table 29, p. 117) and that, at the temperature extremes, survival was higher at higher salinities. Hirota (1959) stated that chlorinity also affected development into the different morphotypes of this species (see p. 131). Ritter-Záhony (1911*a*) considered that salinity could control the transformations between *S. e. elegans* and *S. e. baltica*.

Reeve and Walter (1972*b*) reported that salinity affected survival but not growth or maturation rates in *S. hispida*.

Effects of parasites

Chaetognath parasites are discussed in Chapter 8. Endoparasites include bacteria, protozoans, nematodes, trematodes, and cestodes. Various endoparasites can apparently interfere with the growth of ovaries or testes (Baldasseroni 1914; Russell 1932*a*; Furnestin 1957; Elian 1960; Weinstein 1972; Pearre 1976*b*; Alvariño 1983*c*; Nagawasa *et al.* 1985; Øresland 1986*a*). In some cases, apparent effects on somatic growth—either increased (Pearre 1976*b*) or retarded (Nagasawa *et al.* 1985*b*)—were also noted. Such effects on either size or maturation can thwart attempts to use these criteria to estimate age of individual animals. Fortunately, incidences of parasitism are generally low enough (Alvariño, 1965) for such problems to be unlikely to seriously interfere with the tracking of populations.

Regeneration

Regeneration in chaetognaths has been reviewed by Hyman (1959), Alvariño (1965), and Ghirardelli (1968). Kulmatycki (1918) reported considerable ability of experimentally wounded *Spadella cephaloptera* to regenerate lost parts of the caudal section. However, Ghirardelli (1968) could obtain only limited regeneration of the fins and caudal area behind the seminal vesicles. He noted that it would be unusual to find extensive regenerative ability in any animal group with early germ-line determination.

Several authors have reported field-caught specimens which appeared to be in process of regenerating lost parts of heads (Pierce 1951; Alvariño 1965; Almeida Prado 1968; Ghirardelli 1968; Boltovskoy 1981*b*).

However, Ghirardelli (1968) and King (1979) experimentally amputated heads in *S. enflata* and *S. elegans* (respectively) and none recovered, death following within a few hours. Nagasawa *et al.* (1984) suggested that this may be because wounded chaetognaths succumb to bacterial attack in the laboratory but might heal over in nature, so that failure to obtain regeneration in the laboratory could be artefactual. Furthermore, it has been pointed out (Merinfeld, personal communication) that some other Deuterostoma (*e.g.* Asteroidea, Ophiuroidea) exhibit great regenerative capacity. Thus, while extensive regeneration remains doubtful, it cannot yet be ruled impossible.

On ages of chaetognaths

It has always been difficult to determine actual ages of chaetognaths because they are soft-bodied animals without definable 'landmarks' in their development (unlike the well-catalogued molt sequences of crustaceans). The best we have been able to do is to track sizes-modes (Sameoto 1971) and/or to make use of the ambiguous and rather imprecise maturity stage criteria. The possible effects on lengths or ovary development of parasitism (see p. 69), shrinkage under starvation (see p. 68), and protracted and/or sequential spawning all conspire to make the estimation of ages of individuals and identification of cohorts within chaetognath populations rather precarious; it would be useful to have a less ambiguous indicator of chronological age. Possible candidates might be the 'age pigments' (lipofuscins) which have been experimentally investigated in crustaceans and fish, or pteridine, which has been used less broadly but with similar results (see review by Nicol 1987).

Natural life cycles

Spawning periodicity

There is a large literature on dates of spawning in chaetognaths and not space enough here to do it justice. Alvariño (1965) provides a review of work up to that time. Unfortunately, many studies were unable to distinguish separate broods or cohorts, and so could not confidently delineate life cycles of the individual species. This is true of many warm water species, which often breed over all or a large part of the year (Alvariño 1965), and of populations sampled in areas of strong water mass mixing. For these reasons, much of the most illuminating work has been done in small, semi-isolated bodies of water in highly seasonal areas.

Generation length

It has been considered axiomatic that the generation length of chaetognaths decreases from the poles towards the equator (Owre 1960; Alvariño 1965; Jakobsen 1971), this is generally thought to be a function of temperature (see p. 68). While this is probably true, the evidence is not as strong as it should be. Early speculations were based partly on the gradient of generations found by including Russell's data on *S. elegans* and *S. setosa* in the English Channel off Plymouth, but there have been serious doubts about Russell's interpretation of his data (see p. 71). Studies showing multiple generations in other species may often be subject to the same criticisms (Alvariño 1965) but, because a species is seen to breed year-round (many do in warm areas) this does not necessarily mean it has a short generation length; numerous generations may coexist at different stages of development. Again, we now realize that multiple cohorts may be produced by a single adult generation, leading to greater numbers of perceived generations where spawning seasons are longer. This argues for defining generations by disappearance of older stages rather than appearance of young (Øresland 1985). Some of the better-known and/or more instructive studies are reviewed below:

Sagitta elegans

This is undoubtedly the best-studied species. Weinstein (1972) and Kotori (1979) provide detailed critical reviews of earlier work. Briefly, with updates: Russell (1932a) stated that in the English Channel this species had four to five generations per year, but Jakobsen (1971) pointed out that, because of the considerable advection and mixing of water masses in this area, Russell's specimens may have represented several populations. Pierce (1941) and Khan and Williamson (1970) reported one generation per year in the Irish Sea, whereas Conway and Williams (1986) reported three in the Celtic Sea. Wimpenny (1937) reported three generations per year in the North Sea. Jakobsen (1971) found one generation in Oslofjorden, southern Norway; Sands (1980) reported one in Korsfjorden, western Norway, and Tande (1983) reported a single generation in Balsfjorden, northern Norway, although with possible multiple population modes. Øresland

(1985) reported a single generation in Gullmarsfjorden, as did Båmstedt (1988) for Kosterfjorden, both in southern Sweden.

Huntsman and Reid (1921) reported one generation in the Gulf of St. Lawrence, eastern Canada, although Weinstein (1972) found a biennial life cycle there. Sameoto (1971) reported two generations per year in each of four subpopulations in Saint Margaret's Bay, Nova Scotia—the first serious attempt to distinguish perceived maturity peaks from actual generations. Both Sameoto (1973) and Zo (1973) found two generations per year in the nearby Bedford Basin (three years' data). Redfield and Beale (1940) and Sherman and Shaner (1968) reported one generation per year in the Gulf of Maine, but Clarke *et al.* (1943) suggested that there were two on Georges Bank, adjacent to the south, and thought to be the source for most of the Gulf of Maine population. Clarke and Zinn (1937) felt that there were four breeding periods per year (generations?) immediately south of Cape Cod, USA. Sweatt (1980) reported three generations in the adjacent Rhode Island Sound, as did Deevey (1952) for Block Island Sound, the next Sound to the south. Deevey (1956) thought 'probable' that there were four in Long Island Sound, adjacent to Block Island Sound, but found the population very sparse. However, Tiselius and Peterson (1986) reported a seasonally very abundant population there, with two discernable cohorts during its time of residence. King (1979) reported one or (possibly) two generations per year from a population in Dabob Bay, northwestern USA.

Sagitta elegans comprises three subspecies: *S. elegans elegans*, *S. elegans arctica*, and *S. elegans baltica* (Ritter-Záhony 1911*b*; Fraser 1952). Ussing (1938) and Kramp (1939) found a single generation of *S. e. arctica* around Greenland, and Sameoto (1987) has reported an annual cycle in Baffin Bay. However, Bogorov (1940), Dunbar (1940, 1941, 1962) and McLaren (1966) reported that this species was biennial in arctic waters, although a population of much smaller animals in the semi-landlocked arctic Ogac Lake appeared to be annual (Dunbar 1962; McLaren 1969).

Rózanska (1971) reported three to four generations per year of *S.e. baltica* in two deep basins of the Baltic Sea. However, due to sampling limitations the data were difficult to interpret.

Sagitta setosa

Russell (1932*b*) claimed five to six generations per year in the English Channel, but this conclusion has been criticized by Jakobsen (1971) and Øresland (1983, 1986*b*). Jakobsen (1971) reported two generations per year in Oslofjorden, Norway, and Øresland (1983, 1986*b*) concluded that in the Kattegat and the western English Channel there was a 1-year life span. Wimpenny (1937) and Bainbridge (1963) reported two generations per year in the North Sea, as did Pierce (1941) and Khan and Williamson (1970) in the Irish Sea.

Sagitta enflata

Massuti Oliver (1954) reported two generations per year in the western Mediterranean. Nagasawa and Marumo (1977) report four cohorts (perhaps generations) per year in Sagami Bay, Japan. As a warm water species, it is usually reported to spawn year-round (Owre 1960; Szyper 1976).

Sagitta bipunctata

Massuti Oliver (1954) reported two generations per year in the western Mediterranean.

Sagitta crassa

Hirota (1959) and Murakami (1959) reported three generations per year in the Inland Sea of Japan. Murakami (1966) reported four generations per year in Kasaoka Bay, off the Inland Sea. Morphology varies greatly between generations (see p. 68).

Sagitta gazellae

David (1955) reported a 1-year life cycle in the Antarctic.

Sagitta nagae

Nagasawa and Marumo (1977) suggest six or seven cohorts per year in Sagami Bay, Japan and (1978) suggest that there are probably three generations per year in Suruga Bay, although they acknowledge that interpretation of their data was difficult.

Eukrohnia hamata

Bogorov (1940) reported a biennial life cycle in the Barents Sea., Giskeodegård (1975; in Sands 1980) found a biennial life cycle in Trondheimsfjorden, as did Sands in Korsfjorden, west Norway. Sameoto (1987) has also reported a biennial cycle in Baffin Bay.

Production and production/biomass ratios

Production estimates are necessary for estimating ecological efficiencies and cropping by predators—more usually (and incorrectly) compared to standing crop or biomass—as well as for comparing production differences between populations, areas, or seasons. Production is generally expressed in milligrams of carbon either per unit surface area or volume in a given period; usually daily or annual. Unfortunately, it is often difficult to make these conversions on the basis of information supplied (mean water column depth, period covered by the sampling project), and so not all studies are comparable. Table 7.4 lists a number of studies; where possible I have estimated the appropriate conversion to equivalent units.

Larger populations usually produce more. However, we often want production independent of population size, i.e. the 'specific production', and for this, the usual estimator is the production/biomass ratio, or P/B. This ratio is independent of measurement units but remains dependent on the time period over which it is measured—usually either per day or year. Table 7.4 also lists published estimates of P/B ratios.

Growth efficiency

There have been few measurements made of growth efficiency in chaetognaths. With laboratory culture becoming more common, this situation should improve in the future. Gross growth efficiency was estimated by Reeve (1970a) for *Sagitta hispida* on a dry weight (W_D) or nitrogen weight (W_N) basis as $(W_{i2} - W_{i1})/W_{iD}$, where W_D is the weight of food (diet) consumed in the appropriate units over the course of the experiment. Reeve (1970a) found that, if egg production was included as part of production, the dry weight efficiency of *S. hispida* decreased slightly as a function of chaetognath size but was unaffected by temperature, while nitrogen efficiency was unaffected by size but declined strongly as temperature increased. Overall, immature animals had an average dry weight growth efficiency of 34.5 per cent. This fell to zero during egg production. On a nitrogen basis, however, egg producers averaged about 41 per cent, and the average growth efficiency over the size range studied was 36 per cent.

Nagasawa (1984) reported that the dry weight growth efficiency for *S. crassa* increased during the early life of the animal and then decreased, with a mean of 28 per cent.

Acknowledgements

I would like to thank E. G. Merinfeld of the Department of Oceanography, Dalhousie University, for his broad expertise, and R. J. Conover for a very thorough and helpful review of an earlier version of this text. Perhaps most of all I must acknowledge my enormous debt to Ms Rosemary MacKenzie (Interlibrary Loan Librarian) and the rest of the staff at the Dalhousie University Science Library. Without their unstinting help, this project could never have been launched at all.

Table 7.4 Net production and P/B ratios for chaetognath populations, in miligrams of carbon. Production on water surface or volume basis, as noted, and per day or per year. P/B on daily or annual basis

Species	Area	Period	Production per m³	per m²	P/B	Reference
Daily						
Sagitta enflata	Kaneohe Bay, Hawaii	(1974)				Szyper (1976)
		December–February	—	3.24	—	
		July–August	—	9.00	—	
		(1975) April	—	26.28	—	
		(1974–75) season mean	—	10.55	—	
Sagitta elegans	Ogac Lake, Baffin Island., Canada	(1957)				McLaren (1969)
		June–July	—	0.046	—	
		July	—	0.61	—	
		July–August	—	2.68	—	
		early August	—	2.64	—	
		late August	—	0.46	—	
		early September	—	0.32	—	
		late September	—	0.16	—	
		1957: (season mean)	—	1.02	—	
		1962: (season mean)	—	4.26	—	
	Bedford Basin, Canada	(1967)				Zo (1973)
		June–July	—	0.07	—	
		July	—	4	—	
		July–August	—	3.03	—	
		August–September	—	10.5	—	
		September	—	16.5	—	
		September–October	—	0.67	—	
		1967: (season mean)	—	5.46	—	
	Bering Sea	summer	—	1.4	0.015*	Kotori (1979)
	Long Island Sound, USA	May–June, 1982	0.5	17.50	—	Tiselius and Peterson
		May–June, 1983	0.4	14.20	—	(1986)

Table 7.4 (continued)

Species	Area	Period	Production per m³	per m²	P/B	Reference
Sagitta hispida	Card Sound, Florida USA	January	—	—	0.24	Reeve and Baker (1975)
		February	—	—	0.22	
		March	—	—	0.26	
		April	—	—	0.24	
		May	—	—	0.3	
		June	—	—	0.36	
		July	—	—	0.39	
		August	—	—	0.4	
		September	—	—	0.37	
		October	—	—	0.36	
		November	—	—	0.31	
		December	—	—	0.24	
		mean	—	2	0.31	
		(estimated)	—	4.8		
	Biscayne Bay, Florida by size class (mm)	<2.5	—	—	0.38	
		2.5–3.7	—	—	0.37	
		3.8–4.9	—	—	0.29	
		5.0–6.2	—	—	0.3	
		6.3–7.4	—	—	0.14	
		7.5–8.7	—	—	0.15	
		8.8–9.9	—	—	0.07	
		10.0–11.3	—	—	0.04	
Sagitta setosa	Black Sea 'bathyplankton'		—	—	0.09	Petipa *et al.*, (1970)
	'epiplankton'		—	—	0.18	
		1960	—	—	0.2	Greze (1970)
		1961	—	—	0.21	Zaika (1972)
			—	—	0.31	
		winter	—	—	0.03	Porumb (1982)
		spring	—	—	0.06	
		summer	—	—	0.17	
		autumn	—	—	0.08	

Table 7.4 (continued)

Species	Area	Period	Production per m³	per m²	P/B	Reference
Pterosagitta draco	Hawaii (oceanic)	April	–	–	0.014	Newbury (1978)
		May	–	–	0.023	
		June	–	–	0.06	
'Sagittoidea'	Osaka Bay, Japan	May	0.041	–	0.053	Joh and Uno (1983)
		July	1.143	–	0.11	
		August	1.397	–	0.113	
		October	0.381	–	0.118	
Annual						
Sagitta enflata	Kaneohe Bay, Hawaii		–	3850	–	Szyper (1976)
Sagitta elegans	Ogac Lake, Baffin Island, Canada	1957	–	104.8	1.0–2.1	McLaren (1969)
		1962	–	319.4		
	St Margaret's Bay, Canada	1968–1970				
		subpopulation 1	13.3	–	–	Sameoto (1971)
		subpopulation A	13.3	–	–	
		subpopulation 2	6.8	–	–	
		subpopulation 3	6.8	–	–	
		subpopulation 4	6.8	–	–	
		total	47.1	200	2.0–2.1	
	Beford Basin, Canada	1967–1969	–	579	–	Zo (1973)
	Bedford Basin, Canada	1969–1970	–	2400	2.83	Sameoto (1973)
	Long Island Sound, USA	1982	31.5	1100	–	Tiselius and Peterson (1986)
		1983	24.5	850	–	
Sagitta sp. (probably *S. elegans*)	Kiel Bight, Germany		–	2870	–	Martens (1976)
Sagitta hispida	Card Sound, Florida, USA		–	730	109	Reeve and Baker (1975)
	Biscayne Bay, Florida (est.)		–	1752		

* based on biomass = 27.1 mg W_w/m^3 (Kotori 1976)

8

PARASITISM AND DISEASES IN CHAETOGNATHS

Sachiko Nagasawa

The early records of parasites from chaetognaths (reviewed by Hyman (1959) and Alvariño (1965)) consist mainly of isolated descriptive observations. They include protozoans, nematodes, trematodes and cestodes. Dollfus (1960) catalogued all early records of digenetic trematodes from chaetognaths and attempted to classify them on a modern systematic basis. Rebecq (1965) included lists of trematode cercariae found free in the plankton as well as those found in chaetognaths. Weinstein (1972) examined the chaetognath as host and elucidated the relation between the biology of the host and the dynamics of its parasites. Studies by Pearre (1976b, 1979a) were concerned with the physiological effects of larval trematodes on their hosts and on the host population structures and with parasite-induced morphological or behavioural changes in the chaetognaths.

The major chaetognath parasites are first described, and then the ecological significance of host–parasite relationships and disease in chaetognaths will be considered. By disease, is denoted a demonstrable negative deviation from the normal state of a living organism (Kinne 1980). Here, 'negative deviation' implies a functional or structural impairment, which is quantifiable in terms of a reduction in ecological potential, such as survival, growth, reproduction, energy procurement, stress endurance, and competition.

Periphytes

The microbial colonization sometimes observed on chaetognath body surfaces is a periphytic phenomenon, although the micro-organisms have not been identified and it is not known if they are indeed actual external parasites. Many periphytic bacteria, fungi, and yeast depend on the substratum as a source of organic nutrients. In fact, all marine bacteria, fungi, and blue-green algae are potentially periphytic and many are facultative or part-time periphytes, i.e. they are free-living under some conditions. Many other species are obligate periphytes and depend on a suitable substratum for normal growth (ZoBell 1972). Nagasawa *et al.* (1985a) reported microbial colonization of three different types on the body surfaces of *Sagitta crassa* in Tokyo Bay: branch-like growths (most common); large numbers of filaments (least common); and protuberances (common) (Fig. 8.1(a–d)). Branch-like growths and protuberances were also found along with another peculiar periphyte on *Sagitta helenae* obtained from the Atlantic Ocean. This unidentified periphyte has three spinous processes which appear to be able to elongate (Fig. 8.1(e,f)). Micro-organisms attached to the surfaces of living chaetognaths have been frequently observed (Nagasawa and Nemoto 1984). However, little information is available concerning the ecological aspects of these micro-organisms when compared with the information that has been provided on the association of bacteria with the surfaces of marine copepods (Kaneko and Colwell 1975; Sochard *et al.* 1979; Huq *et al.* 1983; Nagasawa *et al.* 1985a; Nagasawa and Nemoto 1986; Nagasawa 1986a,b; Nagasawa 1987a; Nagasawa and Terazaki 1987; Nagasawa *et al.* 1987; Nagasawa 1988b; Nagasawa 1989b).

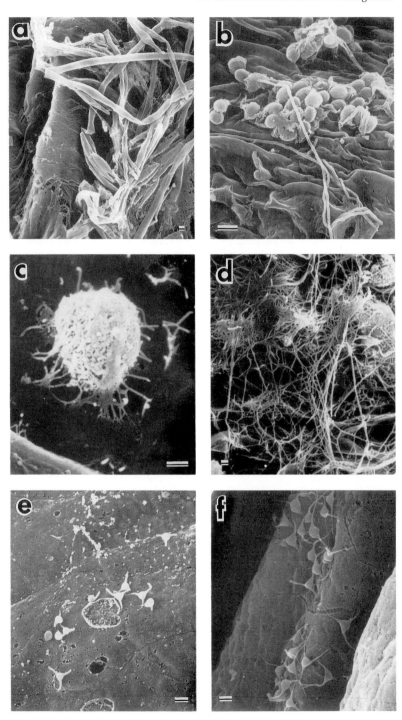

Fig. 8.1 Periphytes of chaetognaths. (a, b) Branch-like growths, (c) a protuberance, (d) filaments, (e, f) three spinous processes. (a) A branch-like growth on the body of *S. crassa*, (b) many egg-shaped and branch-like growths on the body of *S. crassa*, (c) a protuberance found in *S. crassa*, (d) large numbers of filaments found in *S. crassa*, (e) several periphytes with three spinous processes found in *S. helenae*, (f) these spinous processes are longer than those in (e) suggesting that they grow longer on the body surface of *S. helenae*. Scale bars indicate 1 μm (a, c, d, e, f) and 10 μm (b).

The protozoan parasites of the chaetognaths

Infestations by multiple individuals are the rule in the protozoan parasitism.

Dinoflagellates

The ectoparasite on *Sagitta elegans* at Friday Harbor (Washington) is a trophozoite of the parasitic dinoflagellate *Oodinium jordani* and has recently been described by McLean and Nielsen (1989). This new species attaches to the chaetognath fins and the penetrating peduncle causes extensive damage.

Gregarines

Large numbers of gregarines were found in the gut of seven epipelagic species: *Sagitta nagae*, *S. enflata*, *S. pacifica*, *S. ferox*, *S. neglecta*, *S. regularis*, and *Krohnitta pacifica* (Nagasawa and Marumo 1979a). Most belonged to the genus *Lankesteria* but differ from *L. leuckarti* (Shimazu 1979). Living specimens of these gregarines are shown in Figure 8.2 (a, b).

Specimens of *S. enflata* from Bay of Alger (Ramult and Rose 1945) and Bay of Nha Trang (Hamon 1956) harboured *L. leuckarti*. Furnestin (1957) inferred that *S. friderici* and *S. bipunctata* are infected with *L. leuckarti*.

Nagasawa and Marumo's (1979a) studies on the incidence of gregarine infection in Suruga Bay revealed that it ranged from 7 to 56 per cent in *S. regularis*, from 0 to 23 per cent in *S. neglecta*, and from 0 to 50 per cent in *Krohnitta pacifica*. These three species of chaetognath were advected to Japanese waters and their number peaked in September (Marumo and Nagasawa 1973). They were infected more frequently than *S. nagae*, which lives in Suruga Bay, and gregarines were more numerous in their intestines than in the intestines of *S. nagae*. Nagasawa and Marumo (1979a) therefore postulated that the three species of chaetognaths were infected with gregarines before reaching Suruga Bay.

Ciliates

Metaphrya sagittae, which was first recorded by Ikeda (1917), packs dense clusters of large cells into the trunk coelom in *S. enflata* (Ramult and Rose 1945; Furnestin 1957; Nagasawa and Marumo 1979a), *S. minima* (Ghirardelli 1950, 1952; Furnestin 1957), *S. bipunctata* (Massuti 1954), *S. nagae* and *S. pacifica* (Nagasawa and Marumo 1979a), *S. elegans* (Weinstein 1972), *S. tasmanica* (Jarling and Kapp 1985), and *Eukrohnia hamata* (Stadel 1958). The estimated average intensity of infestation was of the order of 100 cells per host (Weinstein 1972). Weinstein, who observed *M. sagittae* in the main body coelom of preserved *S. elegans*, reported that the ciliates were dispersed randomly throughout the coelom. Nagasawa and Marumo (1979a) described how large numbers of *M. sagittae* moved in a line in the trunk coelom of living hosts (Fig. 8.2(c,d)) and showed photographs of random distribution of ciliates.

Weinstein (1972) found an extremely high proportion of the largest *S. elegans* to be infected by *M. sagittae*. The infested specimens had a spent appearance, suggesting the possibility that *M. sagittae* infests the host during or after spawning. According to Weinstein, animals infected by *M. sagittae* had an overall abnormal appearance, both in general morphology and in the gonads of both sexes; the body was also flaccid and swollen. In some specimens the sperm in the tail coelom appeared fragmented, while in the others the tail coelom contained only a clear yellow fluid. Ovaries were considerably shorter than normal for the host's body size and contained no developed ova.

Metaphrya sagittae generally infests less than 1 per cent of the host population at any time and the highest incidence found was 2.6 per cent (Weinstein 1972).

The helminth parasites of the chaetognaths

The helminths are the most frequently encountered parasites of chaetognaths. Larval trematodes and nematodes are reported most often, although cestodes are encountered only rarely. Multiple infestation is exceptional for the helminths.

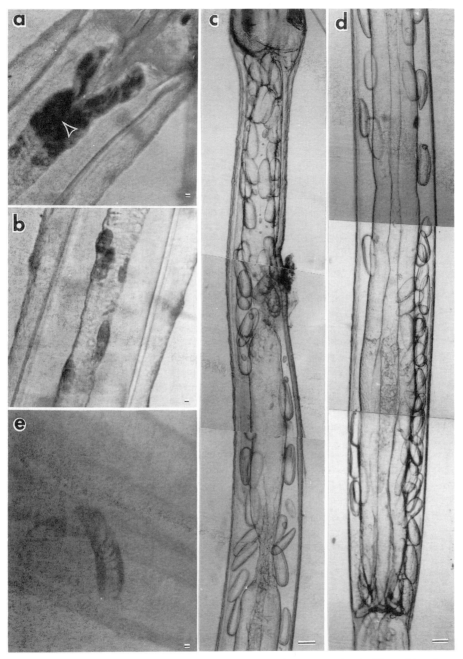

Fig. 8.2 Parasites of chaetognaths. (a, b) Gregarines, *Lankesteria* sp., (c, d) ciliates, *Metaphrya sagittae*, (e) *Monilicaecum* group metacercariae. (a) a large number of gregarines (arrow) are present at the upper part of gut of *S. regularis*, (b) a small number of gregarines are found in part of gut of *S. neglecta*. Ciliates are present in the anterior (c) and posterior (d) parts of *S. nagae*, (e) metacercaria piercing the gut wall of *S. nagae* and coming out of it. Scale bars indicate 10 μm (a, b, e) and 100 μm (c, d). Reproduced, with permission of the publisher, from Nagasawa and Marumo (1979).

Larval trematodes (metacercariae)

Sagitta enflata, *S. ferox*, *S. nagae*, *S. neglecta*, and *S. regularis* obtained from Suruga Bay harboured *Monilicaecum* group metacercariae (Nagasawa and Marumo 1979a). *Sagitta bipunctata*, *S. enflata*, *S. friderici*, *S. minima*, *S. serratodentata*, *S. hexaptera*, *Pterosagitta draco*, and *Spadella* sp. from the Atlantic Ocean were infected with metacercaria of the Family Didymozoidae (Reimer *et al.* 1975), which appear as *Monilicaecum* group metacercariae (Shimazu 1978). In addition to *Monilicaecum* group metacercariae, chaetognaths in Suruga Bay were infected with metacercaria of the *Torticaecum* group in the Family Didymozoidae and of *Tetrochetus* sp. and *Guschanskiana* sp. in the Family Accacoeliidae (Nagasawa and Marumo 1979a). Weinstein (1972) reported that *Hemiurus levinseni* metacercariae were the most common and abundant parasites of *S. elegans* in the Gulf of St. Lawrence. This parasite was always seen in the host's body coelom, and was never found in the digestive tract or in the tail coelom. A metacercaria of *Ectenurus lepidus* was found predominantly in the tail coelom, except in *S. minima* (Jarling and Kapp 1985). There seemed to be no special preference for any particular location within the coelom, although some authors have reported trematodes in the region of the ovaries (Weinstein 1972). My observations of living chaetognaths with metacercaria support this, because metacercariae move around the body coelom actively and freely. In addition, they were found in the head, body coelom and ovaries (Nagasawa and Marumo 1979a).

Weinstein (1972) reported that *H. levinseni* metacercariae were always present, but had a definite seasonal peak during the autumn months. According to him, 4.8 per cent of all chaetognaths examined harboured the trematode; the intensity of infestation varied from one to six trematodes per host and multiple infestations were exceptional. Multiple infestations occurred in some specimens of *S. enflata* from the East China Sea; they harboured at least four *Monilicaecum* group metacercariae in their trunk and tail coelom (Nagasawa and Marumo 1981). Most of the chaetognaths off northwest Africa harboured only one metacercaria of *E. lepidus*, but some animals had two or three (Jarling and Kapp 1985). All the trematodes (mostly metacercariae) were found in the body coelom except for one *Derogenes varicus* and eight *Lecithochirium* specimens (in the tail) and two *Monascus filiformis* (in the gut) (Øresland 1986a).

Larval trematodes of *Tergestia* sp. were found only in the gut of *Sagitta crassa* in Tokyo Bay (Nagasawa and Marumo 1984a). Most infected chaetognaths (88–100 per cent) had one metacercaria; some had more. The percentage of infection by this larval trematode ranged from 0.8 to 4.7 per cent. Specimens of *S. crassa* smaller than 2.5 mm were not infected, but some specimens larger than 2.5 mm were, suggesting that the feeding habit of the host varies and that specimens larger than 2.5 mm are exposed to the risk of *Tergestia* infection (Nagasawa and Marumo 1984a).

High levels of infestation by *H. levinseni* only occurred in *S. elegans* larger than 21 mm (Weinstein 1972).

The nearshore species, *S. friderici*, had the highest incidence of *E. lepidus*, suggesting that infection is most likely to occur in shallow water. These metacercariae are not only larger but also more mature further offshore; they grow while their hosts drift seaward. (Jarling and Kapp 1985). *H. levinseni* metacercariae had no apparent effect on ovary development in their hosts and infested animals possessed the characteristics of normal, healthy chaetognaths at the state of maturity appropriate to their size (Weinstein 1972). Although some mechanical damage by these coelomic parasites was to be expected, no punctures or recently healed wounds were seen on the digestive tract or body wall. Øresland (1986a) also mentioned that no injuries due to parasites were seen. In examining living chaetognaths infected by larval trematodes Nagasawa and Marumo (1979a) reported that *Monilicaecum* group metacercariae usually moved about in the trunk coelom, repeatedly contracting and extending. On one occasion they observed one metacercaria in the process of piercing the gut wall (Fig 8.2 (e)). This suggests that larval trematodes may damage some parts of the host, but it will obviously be difficult to detect this in preserved specimens.

Nagasawa and Marumo (1981) found specimens of *S. enflata*, infected with metacercariae belonging to *Lecithocladium* of the Family Hemiuridae and the *Monilicaecum* group, in the gut contents of the flat fish *Cleisthenes pinetorum*, suggesting that this fish may serve as the definitive host for these parasites. Thus, parasites could be used as a label, much like a radioactive element on a metabolite, to trace the pathways through which chaetognath material travels in the marine food web (Weinstein 1972).

Larval nematodes

Nematodes have been found frequently and in high incidence in chaetognaths. Most have been identified as larval ascaroids; their specific identity is difficult to determine because of the similarity between the larval forms of ascaroid genera.

Only two *Contracaecum*-type larvae were seen in nearly 10 000 *S. elegans* examined in the Gulf of St. Lawrence (Weinstein 1973), suggesting that this host plays a negligible role in the nematode's life cycle in these waters. In contrast, relatively high incidences of nematodes have been reported in some chaetognath populations. Russell (1932*b*) reported that in August 1930 7 per cent of *S. setosa* off Plymouth were infected and Elian (1960) found that 2 per cent of *S. euxina* from Romanian waters were infected. Ass (1961) reported 33 per cent of *Sagitta* sp. in the Black Sea to be infected with larval *Contracaecum* sp. He concluded that *Sagitta* played an important role as an intermediate host in the nematode life cycle. Four of the five infected specimens of *S. friderici* from the Patagonian Shelf were larval nematodes of the *Contracaecum* type (Jarling and Kapp 1985). Among the species of eight parasites of *S. setosa* off Plymouth, Øresland (1986*a*) reported that larvae of the nematode, *Hysterothylacium aduncum* (syn.: *Contracaecum aduncum*, *Thynnascaris adunca*), were the most common (accounting for 56 per cent of all parasites found), although the incidence in *S. setosa* varied from 0 to 7 per cent. All *H. aduncum* were found in the coelom. Some specimens were longer than the trunk of their host and consequently were coiled in the coelom. Where the nematode larva was present, the coelom contained a short ovary. The occurrence of short ovaries is probably due to retardation of growth rather than direct influence by parasites (Øresland 1986*a*). The few short ovaries, together with the low percentage of infection, may indicate that the reproduction of the *S. setosa* population off Plymouth was not affected by parasites in 1982–3 (Øresland 1986*a*). Øresland also inferred, on the basis of previous records, that *S. setosa* harbours *H. aduncum* throughout most of its distribution area.

Some specimens of *S. enflata* in the East China Sea were infected with a larval nematode of the Family Anisakidae *Thynnascaris* (Nagasawa and Marumo 1981). This genus includes many identified species whose adults are found in the stomach or intestine of many fishes.

Larval cestodes (metacestodes)

Since chaetognaths are rarely found with cestodes, they must be considered accidental hosts and probably represent a cul-de-sac for the parasite.

Two tetraphyllidean cestodes of *Scolex pleuronectis*-type larvae were found in *S. elegans* from the Gulf of St. Lawrence (Weinstein 1972). Some larval cestodes were found infecting *Eukrohnia hamata* and *Pterosagitta draco* inhabiting Suruga Bay at 150 m (Nagasawa and Marumo 1979*a*). These metacestodes belonged to the Order Tetraphyllidea, but it was impossible to identify them to family level. Jarling and Kapp (1985) found one *S. tasmanica* off north-west Africa which harboured a larval cestode, while Øresland (1986*a*) reported a cestode larva in the trunk coelom of *S. setosa* off Plymouth.

Aspects of host–parasite relations

Many parasites have evolved to induce morphological or behavioural changes in their intermediate hosts which appear designed to attract the final predatory hosts. There have been a number of instances produced by trematode larvae (Rothschild 1962) and an extensive review of literature describing such effects (Holmes and Bethel 1972). The effect of the parasite on the host population must be expressed in reproductive terms. Damage to the population's reproductive potential is a function of change in fecundity and/or the mortality rate. Parasite-induced excess growth and parasite-induced behavioural change are usually discussed with a possible vector of infection: gigantism and inhibition of ovarian development in chaetognaths as gross physiological effects of larval trematodes; alteration of colour, size, morphology, behaviour or normal habitat as the sum of effects of larval trematodes on the chaetognaths.

A possible vector

Pearre (1976*b*) found that sympatric populations of *Sagitta minima*, *S. friderici* and *S. enflata* had different mean incidences of parasitism by larval hemiurid trematodes. He also showed that the incidence of parasitism increased with the size of the host, both within and between species. This trend is to be expected on the (reasonable) assumptions that larger chaetognaths eat more (Pearre 1974, 1976*a*), are older, and that the parasites are obtained directly or indirectly by feeding (Lebour 1917). Although the routes of infection are not known, *Paracalanus* (the shallowest-living copepod) is a possible vector of the trematode because it was shown to be a food item heavily selected by infected chaetognaths (Pearre 1976*b*). Weinstein (1972) had previously postulated that chaetognaths picked up parasites by feeding on infected copepods, and Nagasawa and Marumo (1984*a*) inferred that the larval trematode of *Tergestia* sp., which was found only in the gut of *S. crassa*, was acquired through predation on copepods infected with this parasite.

Gigantism

The increase in size associated with trematode parasitism is accompanied by a reduction in the size and maturity of the ovaries. In some cases of single trematode infestation, the ovary on the side near the parasite was reduced while the other appeared normal (Furnestin 1957; Pearre 1976*b*). Pearre (1976*b*) discussed a number of possible explanations for the greater lengths of infected chaetognaths and concluded that the most likely explanation was that parasites suppress ovarian development and, directly or indirectly, stimulate somatic growth at the same time. This would mean that parasitized chaetognaths would be both somewhat larger and less mature than the norm of their cohorts. Parasitism should thus lead to lowered fecundity and lengthened generation time. Nagasawa (1985*a*) examined the size composition of normal and abnormal chaetognaths, and those infected with larval trematodes (*Tergestia* sp.). She concluded that the lack of larger abnormal specimens suggested that bacterial infection had caused higher mortality and/or lower growth rate in chaetognath populations. Although no abnormal specimens larger than 9.0 mm were found, specimens 9.0 to 10.9 mm and infected with parasites were present (Nagasawa 1985*a*). The presence of infected larger chaetognaths may be due to gigantism induced by parasites.

Behavioural change

Infestation by larval trematode parasites causes the chaetognath to be more visible and to live nearer the surface than normal (Pearre 1979*a*). The larval trematode appears opaque and whitish, in contrast to the nearly perfect transparency of its host. Together with its greater than normal size, and shallower (better lit) habitat, this should render the parasitized chaetognaths more conspicuous to sight-hunting predators. In consequence, the infected chaetognaths lose one of the major assumed adaptive advantages of vertical migration, the avoidance of visually directed predators (primarily fish) during the day. In other words, a behavioural change caused by parasitism is at work. Although trematodes do not directly kill their chaetognath hosts, they do effectively alter the reproductive potential and susceptibility to death through predation. Trematodes also lower individual reproductive potential by partial parasitic castration and increased generation length (Pearre 1976*b*). Moreover, the excess field mortality of infected chaetognaths is ascribed to a selective predation of these chaetognaths by fish (Pearre 1979*a*), which may also account for the mortality of infected chaetognaths in Tokyo Bay reported by Nagasawa and Marumo (1984*a*).

Diseases of chaetognaths

A biotic disease is the expression of the status of agent–host interactions, and may either be restricted to the individual affected or transmissible from one individual to another. The significance of disease as an ecological principle has received insufficient attention from most biologists (Kinne 1980) and Walford's (1958) statement that the study of diseases is one of the most serious gaps in our knowledge of marine ecology still stands. This is particularly true for the diseases of chaetognaths, which do not have significant importance for human consumption.

Diseases of unknown causes

Two different types of disease in *S. crassa* were observed in animals maintained in the laboratory by Nagasawa and Nemoto (1984). These diseases reduced the survival of the affected chaetognaths by causing them to manifest apparent structural and functional deviations. One individual (S1) suddenly died without showing any symptoms; the entire body became opaque and constricted and exhibited a grotesque appearance due to degeneration of the epithelium (Fig.

8.3(a–d)). The agent is unknown, but Nagasawa and Nemoto (1984) assumed that S1 died of X-disease. A second individual (S2) suddenly lost the head, which was not damaged. The trunk exhibited a slightly unusual appearance, similar to that of S1. The abnormal regions of S2 are considered to be a premonitory symptom of the X-disease of S1.

The ciliary sense organs of a third individual (S3) were especially abnormal, showing the adhesion of cilia (Nagasawa and Nemoto 1984) (Fig. 8.3(e)). Once these sensory organs (Chapter 3) become abnormal, the chaetognath may die of starvation, because they are used in prey detection (Feigenbaum 1978; Nagasawa and Marumo 1978*b*, 1982). While the cause for the development of this abnormality in the ciliary sense organ remains obscure, Nagasawa and Nemoto (1984) attributed it to an unknown disease—not to bacterial infection.

Specimens suffering from X-diseases are yet to be found in the field samples. However, considering the lower frequency of bacterial infection in the sea relative to that in the laboratory, the frequency of X-diseases may be very low in the sea; unhealthy chaetognaths are unlikely to be caught anyway because of rapid death, sinking and decay.

Bacterial infection

Abnormal chaetognaths, which include head-damaged and deformed specimens, are characterized by a reduction in ecological potential (survival, growth, reproduction and competition). Chaetognaths kept in the laboratory, as well as those freshly obtained from the sea, often suffered from bacterial infection which led to damage of the head and other organs (Nagasawa *et al.* 1984). In one instance, the cilia of the ciliary sense organs were not visible on the day before the death of an individual (S4) which resulted from bacterial attack (Nagasawa and Nemoto 1984). Bacterial infection of ciliary sense organs took place in cultivated *S. crassa* at

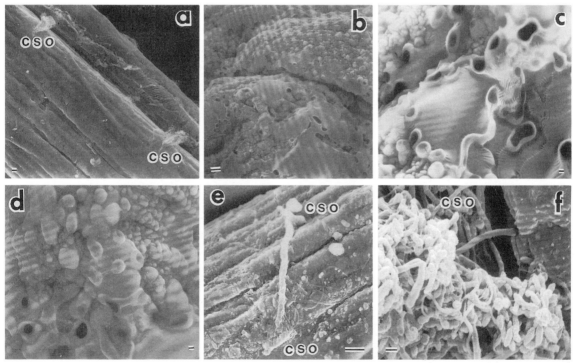

Fig. 8.3 Body surface of chaetognaths which are normal (a) and abnormal (b, c, d) and abnormalities in ciliary sense organs (e, f). (a) *S. ferox*. (b–f) *S. crassa*. The body of the healthy specimen (a) looks smooth and has ciliary sense organs (cso), whereas the body of S1, which suddenly died of an unknown disease, has a rough and grotesque appearance (b, c, d). Two ciliary sense organs are abnormally connected by their cilia, which have become stuck (e). This is also an unknown disease. Bacterial infection of a ciliary sense organ of S4 (f). Scale bars indicate 1 μm (c, d, f) and 10 μm (a, b, e). Reproduced, with permission of the publisher, from Nagasawa and Nemoto (1984).

the rate of 1 per cent (Fig. 8.3(f)). Such bacterial infections, which may have prevented chaetognaths from feeding because of structural and functional deviations, were also observed in *S. helenae* obtained from the Atlantic Ocean.

Large numbers of bacteria were present in head-

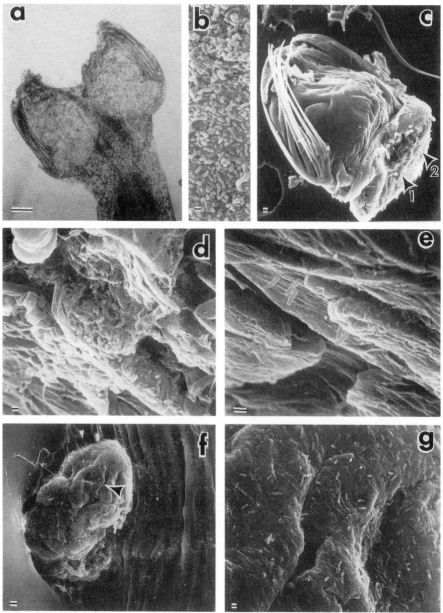

Fig. 8.4 Bacterial infection of chaetognaths. (a) *Sagitta nagae* with head damage; (b) large numbers of bacteria are present on such head-damaged specimens; (c) severed head of deformed *S. crassa*; (d) higher magnification of the part marked '1' in (c)—a large colony of bacteria is present; (e) enlargement of the part marked '2' in (c)—several bacteria are present. (f) An entire view of a tumour-like swelling on the left side of *S. crassa*, (g) enlargement of the part indicated by an arrow in (f). Bacteria are scattered on this outgrowth. Scale bars indicate 1 μm (b, d, e, g), 10 μm (c, f) and 100 μm (a). Reproduced, with permission of the publishers, from Nagasawa (1985a) (c, d, e) and from Nagasawa (1985b) (f, g).

damaged chaetognaths from the laboratory and Suruga Bay (Nagasawa *et al.* 1984) (Fig. 8.4(a, b)). They found growth of *Flexibacter*-like bacteria between the muscle fibres of chaetognaths which had damaged heads.

Abnormal flabby-looking chaetognaths associated with bacteria were found in Tokyo Bay (Fig. 8.4(c–e)) (Nagasawa 1985*a*; Nagasawa *et al.* 1985*b*). In the sample including these deformed chaetognaths was a specimen of *S. crassa* with a pair of tumour-like swellings. With closer observation these turned out to be a massive aggregate of bacteria (Nagasawa 1985*d*) (Fig. 8.4(f,g)). Nagasawa (1985*d*) suggested that tumour-like swellings on *S. crassa* may not be true tumours, since bacterial infection is included in the non-neoplastic lesions accessioned by the Registry of Tumours in Lower Animals (Harshbarger *et al.* 1981).

Bacterial infection in Suruga Bay seems to cause head damage, whereas in Tokyo Bay it causes chaetognaths to become abnormally irregular and flabby (Nagasawa 1985*a*). Thus, bacteria seem to be agents of two different types of deformity in chaetognaths. There was some evidence that deformed chaetognaths in Tokyo Bay have far lower feeding activity and reproductive behaviour than normal, since muscle damage lowered their ability to catch prey and to copulate (Nagasawa 1985*a*). In addition, the absence of larger deformed chaetognaths in Tokyo Bay seems to be the result of their high mortality and slower growth rates, due to bacterial infection.

The percentage of the chaetognath population suffering from bacterial infection was roughly 10 per cent both in Suruga Bay and Tokyo Bay (Nagasawa 1985*a*); these animals are supposed to decay quickly due to bacterial growth. This value is higher than that of animals infected with parasites. It is obvious that bacterial infection in chaetognaths plays an important role in Japanese waters and is of greater ecological significance than metazoan parasitism.

Conclusion

The effect of parasitism and diseases on the population dynamics of chaetognaths is still very poorly known. One of the parameters necessary for understanding the dynamics is mortality—the sum of predation and disease. The latter is generally overlooked in studies of population dynamics because of the lack of information about the effect of parasitic organisms on mortality and/or reproduction.

The term parasitism refers to an intimate coexistence of heterospecific organisms in which one obtains benefits at the expense of the other, ultimately often inflicting demonstrable negative effects on the other. Hosts may react to parasitism at all levels: from the chemical (immunological) and physiochemical (tissue damage and inflammations) to the population (castration and enhanced mortality). To date little information is available concerning the immunological and physiochemical reactions of chaetognaths to parasitic organisms. Immunological techniques are indispensable for testing the reinfection in cultured chaetognaths of bacteria isolated from the infected chaetognaths. Nevertheless, identification of pathogenic bacteria is equally important. Information concerning castration and enhanced mortality in chaetognath populations is still largely lacking.

Parasites which are transferred through the food chain can be used as labels in search for predator–prey relationship, an example of this nature was cited on page 80. However, to make effective use of parasites as indicators or labels one first has to understand the details of the life cycle, the host specificity of different stages, and the factors responsible for the geographic distribution and changes in incidence and intensity.

The routes of parasitic infection, the mechanism of bacterial infection, causes for abnormality in the ciliary sense organs, the agents of X-diseases, and the effect of periphytes on chaetognaths are all topics that merit further attention.

Acknowledgements

I thank Professor Ju-shey Ho, Department of Biology, California State University for constructive comments on the manuscript.

9

DISTRIBUTION PATTERNS IN CHAETOGNATHA

A C Pierrot-Bults and V R Nair

Introduction

Chaetognatha are holoplanktonic animals, except for the genera *Spadella*, *Bathyspadella* and *Krohnittella*. They are found in all oceans and seas, from coast to coast and from the surface to the bottom; they have even invaded estuarine habitats.

This chapter discusses only general distribution patterns and illustrates these patterns with selected species.

The factors that influence distribution patterns are present day water circulation, physiochemical and ecological parameters (temperature, prey abundance, behaviour of the animals themselves, and competing organisms), and historical events like past water circulation patterns, the morphology of ocean basins and the evolutionary history of the group. The first factors are to a large extent covered by the so-called water mass theory. The gross pattern of ocean circulation provides a set of quite distinct biological provinces as a result of climate and vertical circulation differences (Reid *et al.* 1978). The oceans are divided into cyclonic and anticyclonic gyres, with boundary currents and transitional zones in between. These transitional zones are the richest in species. For example, the subtropical convergence is a zone of congregation of tropical and cold water species. For chaetognaths the east and west Pacific boundary currents are the most diverse in species composition (Reid *et al.* 1978). Beklemishev (1971) divided the pelagic communities into cyclical, terminal, terminal/cyclical, and ice/neritec (Fig. 9.1). For general discussions about zooplankton distributions, see Reid *et al.* (1978); Van der Spoel and Pierrot-Bults (1979*a*); Boltovskoy (1981*a*); Van der Spoel and Heyman (1983); Pierrot-Bults *et al.* (1986).

The different water masses for the Atlantic Ocean, the Indian Ocean, and the Pacific Ocean are shown in Figure 9.2 (*a–c*).

The Pacific Ocean shows the greatest complexity and

the Indian Ocean the least in the amount of different water masses (Van der Spoel and Heyman 1983). Watermasses are horizontally and vertically defined, but species distribution seems to be more confined by the climatic zone than by the water mass *per se*. Diurnal vertical migrating populations cross different water masses every day, and thus encounter daily a wider range of environmental differences vertically than they would encounter horizontally in their range.

The historical component is difficult to study as we do not have a fossil record of chaetognaths. However, phylogenetic systematics, area-cladograms and knowledge about palaeo-oceanographic circulation patterns and palaeoclimate can provide insights into the history and phylogeny of the Chaetognatha and of pelagic distribution patterns (Pierrot-Bults and Van der Spoel 1979; Van der Spoel and Pierrot-Bults 1979*a*; Van der Spoel 1983; Van der Spoel and Heyman 1983).

The range of a species is another point of difficulty and should not be defined merely by presence or absence but also by numerical abundance and breeding. The area where sterile expatriates are found does not belong to the proper range of a species. Since knowledge of this is very scarce it is not yet possible to distinguish between the proper range and recordings of sterile expatriates.

The ocean is three-dimensional and ranges of oceanic species should be given in three dimensions. The most applied vertical division is the epipelagic, from 0–200 m, the mesopelagic from 200–1000 m (subdivided into the shallow mesopelagic from 200–600 m, and the deep mesopelagic from 600–1000 m) and the bathypelagic, which is deeper than 1000 m. These vast ranges give rise to horizontally and vertically separated communities and species (David 1963; Pierrot-Bults 1975*a*; 1979, 1982; and see Fig. 9.9). On the other hand, vertical migration might in some cases prevent actual speciation

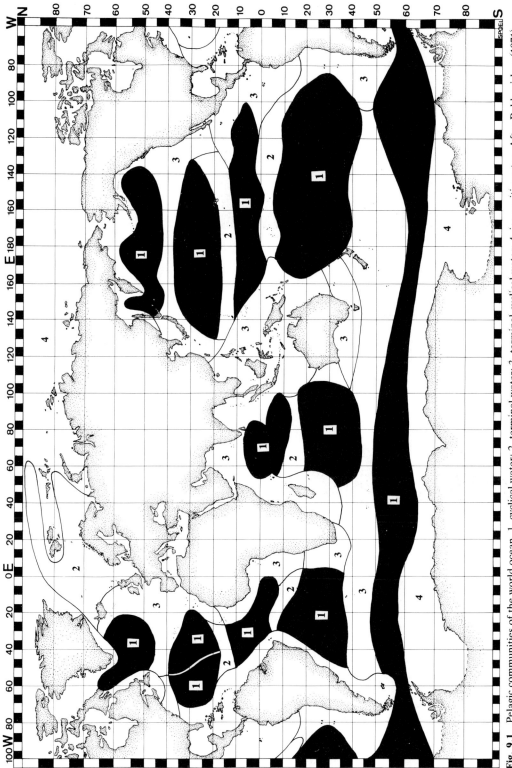

Fig. 9.1 Pelagic communities of the world ocean. 1, cyclical water; 2, terminal water; 3, terminal-cyclical water; 4, ice-neritic water. After Beklemishev (1971).

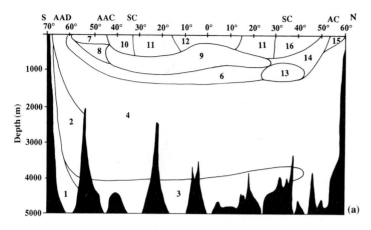

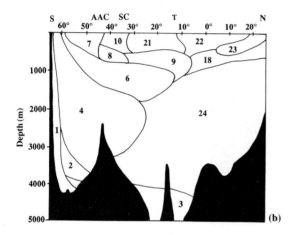

Fig. 9.2 A diagram of the watermasses of (a) the Atlantic Ocean, (b) the Indian Ocean, (c) the Pacific Ocean. 1, Antarctic slope water; 2, circum Antarctic deep water; 3, Antarctic bottom water; 4, Atlantic deep water; 5, Atlantic and Indian deep water; 6, Antarctic intermediate water; 7, Antarctic surface water; 8, intermediate mixed Antarctic water; 9, subantarctic intermediate water; 10, subantarctic surface water; 11, subtropical water; 12, tropical water; 13, Mediterranean water; 14, north Atlantic intermediate mixed water; 15, Arctic water; 16, north Atlantic surface water; 17, north Pacific deep water; 18, tropical intermediate water; 19, north Pacific intermediate water; 20, north Pacific surface water; 21, central Indian water; 22, north Indian water; 23, Red Sea and Persian Gulf water; 24, north Indian deep water. AAC, Antarctic convergence; AC, Arctic Convergence; AAD, Antarctic divergence; SC, subtropical convergence; T, thermal front. After Van der Spoel and Heyman (1983).

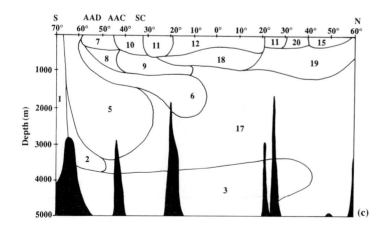

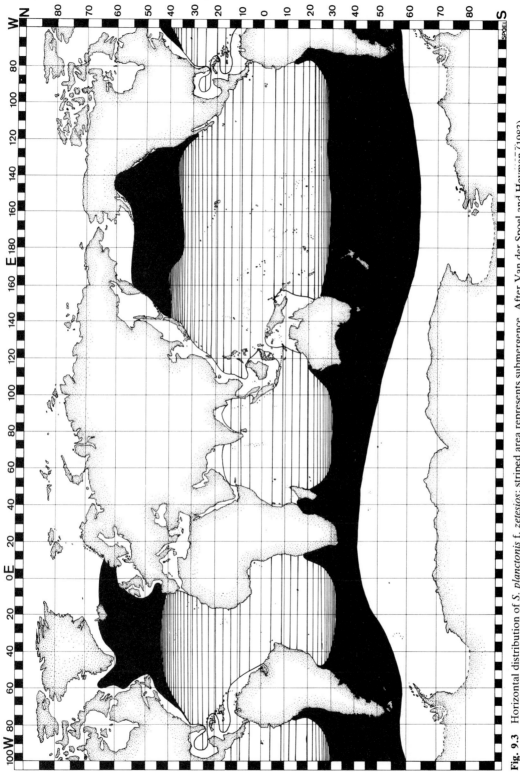

Fig. 9.3 Horizontal distribution of *S. planctonis* f. *zetesios*; striped area represents submergence. After Van der Spoel and Heyman (1983).

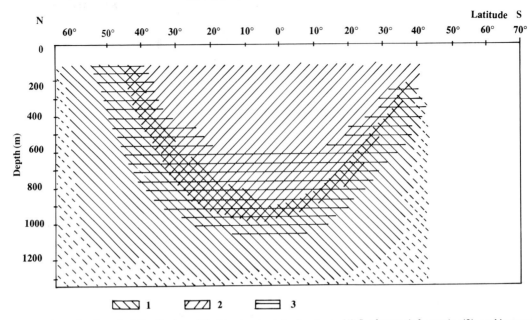

Fig. 9.4 Horizontal and vertical distribution of *Sagitta planctonis* f. *planctonis*, (1) *S. planctonis* f. *zetesios* (2), and intermediate (3). After Pierrot-Bults (1975).

by causing enough mixing of the populations to make at least some gene flow possible (David 1961).

Quantitatively, maximum population density of chaetognaths is confined to the photic zone (Pierrot-Bults 1982). In the tropics and subtropics the epipelagic is a warm biotope and the water column is usually stratified. At higher latitudes there is no permanent stratification and the division between epipelagic and mesopelagic fauna is often lost. Tropical submergence is shown when taxa that live in mesopelagic to bathy-pelagic layers in the (sub)tropics are epipelagic in the higher latitudes, an example is *Sagitta planctonis zete-*

sios Fowler, 1905 (Figs. 9.3 and 9.4) and *Eukrohnia hamata* (Möbius, 1875) (Fig. 9.5). The border between stratified and non-stratified water is not only related to subtropical submergence in some species but also acts as a barrier to the distribution of species in the North Atlantic at about 43°N (Pierrot-Bults 1976a).

The latitudinal shift in bathymetric distribution of some chaetognath species (Fig. 9.6 a–e) shows the emergence of mesopelagic and bathypelagic species to the upper strata towards the subtropical convergence in the Indian Ocean (Nair 1978).

The mesopelagic is characterized by gradually dimin-

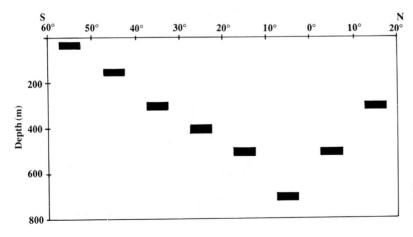

Fig. 9.5 Horizontal and vertical distribution of *Eukrohnia hamata* in the Atlantic Ocean. After Thiel (1938).

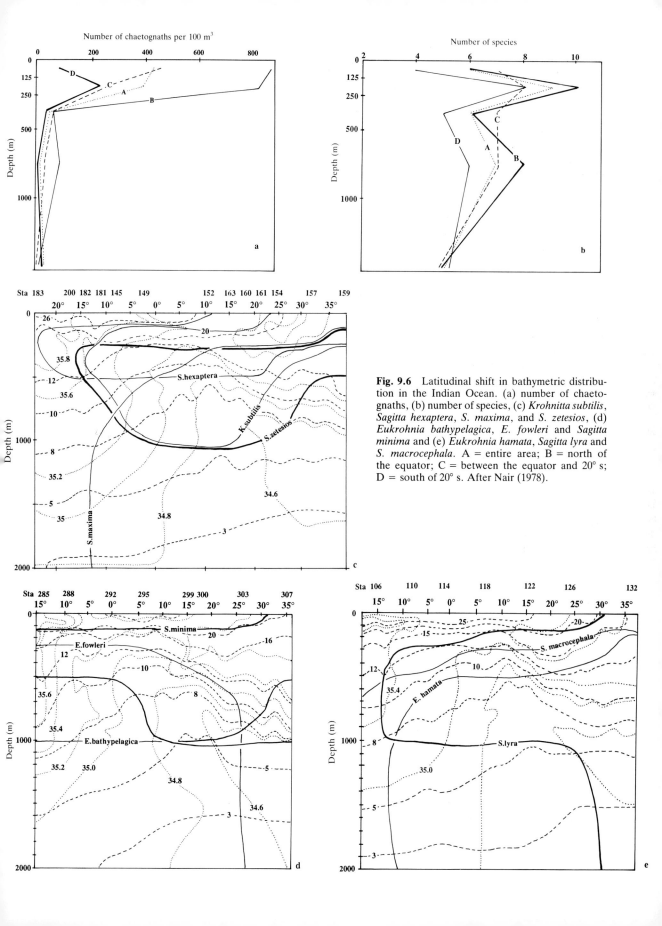

Fig. 9.6 Latitudinal shift in bathymetric distribution in the Indian Ocean. (a) number of chaetognaths, (b) number of species, (c) *Krohnitta subtilis*, *Sagitta hexaptera*, *S. maxima*, and *S. zetesios*, (d) *Eukrohnia bathypelagica*, *E. fowleri* and *Sagitta minima* and (e) *Eukrohnia hamata*, *Sagitta lyra* and *S. macrocephala*. A = entire area; B = north of the equator; C = between the equator and 20° s; D = south of 20° s. After Nair (1978).

ishing light. Most organisms living habitually in these layers perform diurnal vertical migration, living at mesopelagic depths in the daytime and in shallower water at night. Various hypotheses are put forward to explain vertical migrations, for example, the availability of food in the superficial layers and the advantage of lower metabolism at the deeper layers; avoidance of visual predation; horizontal dispersion and transport and breeding migrations (McLaren 1963; Pearre 1973, 1974, 1979b; Sameoto 1973; Longhurst 1976; Angel 1986; Schalk 1988). There are very few records of bathypelagic species performing diurnal vertical migrations; these are generally related to the light cycle (Longhurst 1976; Angel 1986).

Species diversity of chaetognaths gradually decreases from the lower epipelagic to the bathypelagic layers. Because of the overlap of epipelagic and mesopelagic species, maximum species diversity often lies at about 150–250 m, a comparable increase is seen at 800–1000 m as a result of extension of bathypelagic species in the

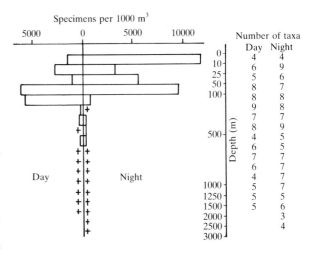

Fig. 9.7 Bathymetric distribution of the number of specimens per 1000 m³ and the number of taxa in the north-west Atlantic. After Pierrot-Bults (1982).

Table 9.1   Species observed at different strata in the Northern Indian Ocean (north of 10° S Lat)

Depth of haul (m)	Water masses found at particular stratum in the western side (west of 80°E long.)	Water masses found at particular stratum in the eastern side (east of 80°E long.)	Species
200–0	Arabian Sea surface water	Bay of Bengal surface water	*K. pacifica, K. subtilis, P. draco, S. bedoti, S. bipunctata, S. enflata, S. ferox, S. hexaptera, S. minima, S. neglecta, S. pacifica, S. pulchra, S. regularis, S. robusta.*
500–200	Persian Gulf water, Red Sea water	Bay of Bengal sub-surface water, Persian Gulf water	*K. subtilis, P. draco, S. bipunctata, S. decipiens, S. enflata, S. ferox, S. hexaptera, S. lyra, S. macrocephala*, S. maxima, S. minima, S. pacifica, S. regularis, S. zetesios.*
1000–500	Arabian Sea intermediate water, Equatorial intermediate water, Pacific water, Antarctic intermediate water	Bay of Bengal sub-surface water	*E. fowleri, E. hamata, S. decipiens, S. lyra, S. macrocephala, S. maxima, S. zetesios.*
2000–1000†	Arabian Sea intermediate water, Arabian Sea deep water		*E. bathypelagica, E. fowleri, E. hamata, S. decipiens, S. lyra, S. macrocephala, S. maxima, S. zetesios.*

* Found only in the northwestern Indian Ocean.

† No data available for this stratum from the northeastern Indian Ocean.

Table 9.2 Species observed at different strata in the central and southern Indian Ocean (between 10° S and 38° S lat.) and at the subtropical convergence (south of 38° S lat)

Depth of haul (m)	Central and Southern Indian Ocean		Subtropical convergence	
	Water masses found at particular stratum	Species	Water masses found at particular stratum	Species
200–0	Surface water	*K. subtilis, P. draco, S. bipunctata, S. enflata, S. ferox, S. hexaptera, S. minima, S. pacifica, S. regularis*	Antarctic water	*S. decipiens, S. lyra, S. maxima, S. serratodentata S. tasmanica*
500–200	Subsurface water	*E. hamata, E. fowleri, K. subtilis, P. draco, S. bipunctata, S. decipiens, S. enflata, S. lyra, S. macrocephala, S. maxima, S. minima, S. pacifica S. serratodentata, S. tasmanica, S. zetesios*	Deep Water	*E. bathypelagica, E. hamata, S. decipiens, S. lyra, S. maxima, S. serratodentata S. tasmanica, S. zetesios*
1000–500	Antarctic water, Pacific water	*E. bathypelagica, E. hamata, E. fowleri, S. decipiens, S. lyra, S. macrocephala, S. tasmanica, S. zetesios*	Deep Water	*E. bathypelagica, E. hamata S. decipiens, S. gazellae, S. lyra, S. macrocephala, S. maxima, S. zetesios*
2000–1000	Deep water	*E. bathypelagica, E. fowleri, E. hamata, S. macrocephala, S. tasmanica, S. zetesios*	Deep Water	*E. bathypelagica, E. hamata S. gazellae, S. macrocephala, S. maxima, S. zetesios*

mesopelagic domain (Nair 1978 (Indian Ocean); Pierrot-Bults 1982 (North Atlantic), (Fig. 9.7 and Tables 9.1 and 9.2)). Species diversity generally shows an increase from high to low latitudes, as in most other animal groups (Nair 1976; Pierrot-Bults 1976a; Nair and Madhupratap 1984).

The energy flux in the ocean is a vertical one. The deeper-living organisms are dependent on surface production and thus in the mesopelagic distributions the climate and distribution of surface parameters is still identifiable. The mechanisms determining bathypelagic distributions seem to be different from those which govern the epipelagic and mesopelagic ones (Fasham and Angel 1975; Angel 1979).

A feature of oceanic species is wide, continuous distribution. In these wide ranges geographic, infras-pecific variation due to different selective pressures in the various parts of the range is frequently present (Van der Spoel 1973; Pierrot-Bults 1975a; Pierrot-Bults and Van der Spoel 1979). The exact taxonomic status of some chaetognath taxa is difficult to determine. Research on *Sagitta crassa* Tokioka, 1938, (Murakami 1959; Tokioka 1974a) has shown that, as a result of changing environmental conditions, considerable morphological differences occur in successive generations in the same population. The same phenomenon is described for *S. tasmanica* Thomson, 1947 from the south-east Pacific where, due to change of temperature and salinity, a gradual change to the *selkirki* form described by Fagetti (1958) as a species occurs (Fagetti 1958; 1968). The same occurs in the north-west Atlantic when *S. tasmanica* becomes trapped into a warm-core

gulf stream ring (Fig. 9.8). The populations of *S. setosa*
Müller 1847 show differences in the different parts of
the range. *S. batava* Biersteker and Van der Spoel,
1966 from the Oosterschelde is conspecific with *S.
setosa*. *Sagitta euxina* Moltschanoff, 1909 from the
Black Sea, described as a close relative of *S. setosa*,
might prove not to be a valid species when populations
of the whole range of *S. setosa* are studied (Pierrot-
Bults 1976*a*).

It is to be expected that detailed variation studies
will reveal more of these differences. According to
Gibbs (1986) we should be suspicious of the many so-
called species that have circumglobal distributions, as
they might prove to consist either of different species
or infraspecific taxa. Studies on copepods show that
there exist genetic differences between populations
(Bucklin and Marcus 1985; Carillo *et al.* 1974).

For the same reason, species with disjunct distributions like *S. tenuis*, *S. tasmanica* and *S. bierii* need
further study to see whether the disjunct parts of the
range are inhabited by populations with different
genetic structures. Different species and higher taxonomic groups have different tolerances to transitions
and change in environmental circumstances. This is
reflected in the distributions of the different groups of
organisms. A survey of faunal boundaries of ranges of

Fig. 9.8 (A) Relation between size in mm (x-axis) and number of posterior teeth (y-
axis) in *Sagitta tasmanica* from the north-west Atlantic, (B) *Sagitta tasmanica* (*selkirki*
form) from the north-west Atlantic Gulf Stream warm core ring. 1, *Sagitta tasmanica*
(*selkirki*) from the south-east Pacific; 2, *Sagitta tasmanica* (*selkirki*) from the north-
west Atlantic; 3, *Sagitta tasmanica* from the north-east Atlantic; 4, *Sagitta tasmanica*
from the south-east Atlantic; 5, *Sagitta tasmanica* from the north-west Atlantic.

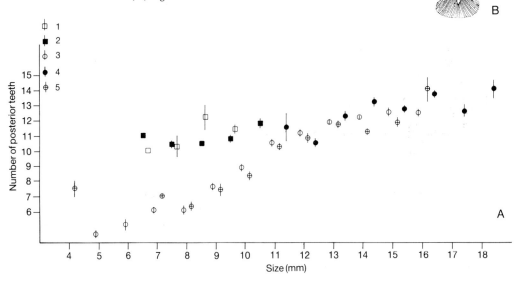

different groups of pelagic organisms shows a general correlation of distribution patterns with water masses, but the precise positions of the distribution boundaries show a gradual change from less tolerant to more tolerant taxa (Van der Spoel and Heyman 1983). Both horizontal and vertical distributions of different species seem to be spatially separated. Figure 9.9 shows the horizontal and vertical distribution of 21 species in the southern Indian Ocean. It is clear that, although there is considerable overlap, the main distribution of each species is different from that of the others (David

1963). The same is seen in the vertical distribution of 18 taxa in the North Atlantic, there is a spatial difference in the distribution of the taxa (Fig. 9.10) (Pierrot-Bults 1982).

The list of species discussed in this paper, along with their distributional ranges in the world oceans, is presented in Table 9.3. These species are chosen because they are considered to be representative of certain patterns of distribution. For a more complete list of chaetognath species see Chapter 11.

Types of distribution

The subtropical gyral regions within the ocean circulation system show the greatest stability in environmental parameters and in stratification of the water column.

Frost and Fleminger (1968) and Fleminger and Hulsemann (1973) recognized latitudinal patterns in the distribution of copepods. They found two main patterns; species occurring from 40°N–40°S and species occurring from 30°N–30°S. The first group showed monotypic species with a circumglobal distribution, the latter showed provincialism with different taxa in the Indo-Pacific and in the Atlantic. Pierrot-Bults (1976a)

compared chaetognath and other holoplanktonic distribution patterns. North–south and east–west discontinuities and ocean connections are important in determining the planktonic distribution patterns, of which latitudinal belt-shaped patterns are most dominant. Van der Spoel (1983) and Van der Spoel and Heyman (1983) divided pelagic distribution patterns further into belt-shaped and non-belt-shaped patterns.

In this paper the above patterns are compared with the distribution patterns of chaetognaths. The distribution maps are reconstructed after Tokioka (1952,

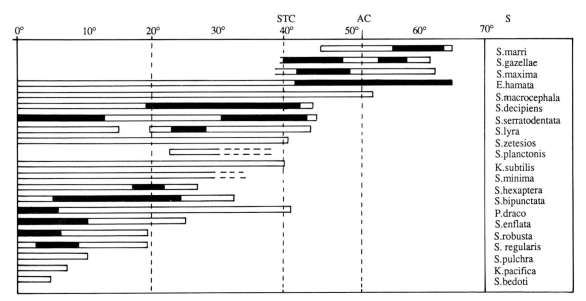

Fig. 9.9 Distribution of 21 chaetognath species in the Indian Ocean. AC = Atlantic convergence, STC=subtropical convergence, S = south. After David (1963).

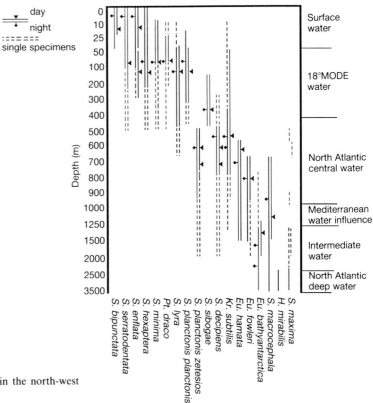

Fig. 9.10 Vertical distribution of 18 taxa in the north-west Atlantic. After Pierrot-Bults (1982).

1959, 1962, 1979); Hida (1957); Bieri (1959); Alvariño (1964, 1969, 1974); Nair and Rao (1973); Pierrot-Bults (1974, 1975a, 1982); Nair (1978); Furnestin (1979); Rao (1979); and Boltovskoy (1981a).

Belt-shaped patterns

Belt-shaped patterns follow the latitudinal or climatological belts and can be divided into wide ranges, and warm water and cold water distributions.

Wide (±70°N–±65°S) ranges -(epipelagic/mesopelagic species)

There is no epipelagic chaetognath species with a very wide, belt-shaped distribution.

Mesopelagic species with a wide, belt-shaped distribution (*Eukrohnia hamata* (Fig. 9.5) and *Sagitta planctonis zetesios* (Figs. 9.3 and 9.4) show (sub)tropical submergence. *Sagitta planctonis zetesios* is replaced in the epipelagic and shallow-mesopelagic of the lower

(<40°) latitudes by its close relative *S. planctonis planctonis* Steinhaus, 1896 (Pierrot-Bults 1975a). Recently, a bathypelagic species belonging to this group has been found in the Atlantic (Chidgey, personal communication). *Eukrohnia fowleri* Ritter-Záhony, 1909 is a mesopelagic species without submergence. Because of competition, its vertical distribution might be influenced by the distribution of *E. hamata* (Pierrot-Bults 1982). In the northern Indian Ocean (north of 20°S) *E. fowleri* dominates the bathypelagic realm but it is replaced by *E. hamata* in subtropical waters (Nair 1978).

Warm water (±40°N–±40°S) ranges (epipelagic/ mesopelagic species)

Many epipelagic chaetognaths show a tropical–subtropical range (from 40°N–40°S) broad enough to enable the species to have a circumglobal distribution and maintain gene flow around South Africa (Fleminger and Hulsemann 1973; Pierrot-Bults 1976a). Examples

Table 9.3 List of species and their distribution in the oceans (species of very restricted distribution excluded)

Species	Ocean		
	Pacific	Indian	Atlantic
Neritic			
S. bombayensis		x	
S. crassa f. crassa	x		
S. crassa f. naikaiensis	x		
S. crassa f. tumida	x		
S. bedoti f. litoralis	x		
S. johorensis	x		
S. demipenna	x		
S. tropica	x	x	
S. oceania	x	x	
S. pulchra	x	x	
S. tenuis	x	x	x
S. friderici	x		x
S. hispida			x
S. helenae			x
S. setosa			x
Epipelagic			
S. bedoti f. bedoti	x	x	
S. bedoti f. minor	x	x	
S. ferox	x	x	
S. neglecta	x	x	
S. regularis	x	x	
S. robusta	x	x	
S. pacifica	x	x	
S. pseudoserratodentata	x		
S. serratodentata serratodentata		x	x
S. serratodentata atlantica	x		
S. tasmanica	x	x	x
S. bierii	x		x
S. bipunctata	x	x	x
S. minima	x	x	x
S. enflata	x	x	x
Pt. draco	x	x	x
Mesopelagic/*cold water			
S. lyra	x	x	x
S. hexaptera	x	x	x
S. decipiens	x	x	x
S. sibogae	x	x	x
S. planctonis f. planctonis	x	x	x
S. planctonis f. zetesios	x	x	x
K. pacifica	x	x	x
K. subtilis	x	x	x
K. mutabbii			x
E. hamata	x	x	x
E. fowleri	x	x	x
*S. gazellae	x	x	x
*S. marri	x	x	x
Bathypelagic			
S. maxima	x	x	x
S. macrocephala	x	x	x
E. bathypelagica	x	x	x
E. bathyantarctica	x	x	x
H. mirabilis	x	x	x

of this type of distribution are *Pterosagitta draco* (Krohn, 1853), (Fig. 9.11), *S. minima* (Grassi, 1881) and *S. enflata* (Grassi, 1883). Whether the monotypy in these species is due to gene flow or is a result of homeostasis is not known. The possibility exists that future research on the different populations of widespread so-called monotypic species will reveal different taxa (Pierrot-Bults 1976a; Gibbs 1986). *Sagitta bipunctata* Quoy and Gaimard, 1827 is a species with a circumglobal 40°N–40°S distribution. It is most abundant in the central water masses, is abundant in the tropical belt of the Indian Ocean (Nair 1977), and is rare or absent in the tropics in the Pacific (Bieri 1959). Casanova and Andreu (1989) did not find it abundant in the tropical Indian Ocean and the abundance pattern in the Atlantic is not clear; it might show the same pattern as the pteropod *Limacina lesueuri* (d'Orbigny, 1938) (Fig. 9.12). *Sagitta lyra* (Krohn, 1853, *S. hexaptera* (d'Orbigny, 1851), *S. sibogae* (Fowler, 1906), non-*S. decipiens*, (Pierrot-Bults, 1979), and *Krohnitta subtilis* (Grassi, 1881) are examples of shallow, mesopelagic belt-shaped distributions. *Sagitta decipiens* Fowler, 1905 is a shallow to deep mesopelagic species with a latitudinal wide distribution. Unfortunately, most authors treat *S. sibogae* and *S. decipiens* as one species and so it is difficult to tell the exact distribution of either. Tokioka (1959) described *S. neodecipiens* (= *S. decipiens*) from the North Pacific Equapac and Shellback expeditions in samples from 400 m to the surface.

The two representatives of the *S. serratodentata* group, *S. serratodentata serratodentata* Krohn, 1853 and *S. pacifica* Tokioka, 1940, have a 40°N–40°S distribution, but they show the non-circumglobal tropical pattern as discussed below (Pierrot-Bults 1974) (Fig. 9.13).

Warm water (±30°N–±30°S) ranges

The species with a restricted tropical range cannot maintain gene flow because the land masses act as physical barriers. According to Fleminger and Hulsemann (1973) this category is characterized by non-circumglobal distributions and different close relatives in the different oceans, usually an Indo-Pacific and an Atlantic species. An exception to this is the epipelagic–shallow mesopelagic *Krohnitta pacifica* (Aida, 1897) (Fig. 9.14) which shows no recognizable differences between the Atlantic and the Indo-Pacific populations, although it might consist of two different species as proposed by Alvariño (1969) who described the Atlantic populations as *Kr. mutabbii*.

Cold water ranges

Chaetognath species occurring at latitudes higher than 40°S are usually circumantarctic. Such is the case for *Sagitta tasmanica, S. gazellae* Ritter-Záhony 1909, *S. maxima* Conant, 1896, and *S. marri* (David 1956) (Fig. 9.15).

More provincialism is seen in the northern temperate zone. *S. tasmanica* occurs in the North Atlantic but not in the North Pacific. *S. elegans* Verrill, 1873 occurs in both oceans (Fig. 9.16). It shows infraspecific variation, with different subspecies in the Arctic (*S. elegans arctica* Aurivillus, 1896) in the subarctic (*S. elegans elegans*) and in the Baltic (*S. elegans baltica* Ritter-Záhony, 1911). Whether the northern and southern hemisphere populations of *S. tasmanica* belong to the same species or not is still an unsolved question (See Fig. 9.15). *S. maxima* has a cold water distribution. It shows tropical submergence but a deep tropical connection between the northern and the southern hemisphere populations has not been found. Differences between the northern and southern populations have been mentioned (David 1963; Pierrot-Bults 1976a) and Bieri (personal communication) proposes a different species for the Antarctic populations.

Patterns derived from belt-shaped patterns

Central water patterns

The only chaetognath species confined to central waters is *S. pseudoserratodentata* Tokioka, 1939 which is found in the North Pacific central water mass (Bieri 1959).

A species with a circumglobal 40°N–40°S distribution, but most abundant in the central water masses of the Pacific and rare or absent in the tropics is *S. bipunctata* (Bieri 1959). In the Indian Ocean it is distributed in the tropics (Nair 1977) but David (1963) records it more abundant from 5–25°S than from the equatorial region. There are no records for abundance in the Atlantic; it might exhibit a distribution pattern similar to that shown in Figure 9.12.

According to Beklemishev (1981) the Pacific equatorial and central waters differ more from each other than is the case in the other oceans, so a more even distribution in the Indian and Atlantic Oceans could be expected. However, the pattern of *S. bipunctata* does not conform with the patterns of other planktonic species, such as the euphausiid *Stylocheiron suhmi* G. O. Sars, 1883, which does not extend into the northern Indian Ocean (Van der Spoel and Pierrot-Bults 1979a; Beklemishev 1981; Van der Spoel and Heyman 1983).

Interocean connections

Interocean connections are possible for cold water species in the Arctic. *S. elegans* and *E. hamata* are distributed in both the North Pacific and the North Atlantic. In the southern hemisphere the cold water and subtropical distributions are circumglobal. The neritic species *S. friderici* Ritter-Záhony, 1911 is distributed in the east Pacific and east Atlantic, including the Mediterranean and is sparsely distributed in the Gulf of Mexico (Fig. 9.17). *S. tenuis* Conant, 1896 shows the same distribution but has a greater representation in the Gulf of Mexico and a narrower distribution in the east Pacific (Fig. 9.18) (Tokioka 1979). This pattern possibly reflects the Panama connection, present until 3.5 mya. Tokioka (1979) records that *S. tenuis* inhabits warmer and more saline water than *S. friderici* and states that *S. tenuis* is the inlet water form of *S. friderici*, but most authors treat both species as valid species. Alvariño (1961) considers the east Pacific populations of *S. friderici* to be a different species, *S. euneritica*.

The exchange between the Indian and the Pacific Ocean through the Indo-Malayan archipelago is predominantly from the Pacific towards the Indian Ocean (Gordon 1986). In the Banda Sea, strong and deep vertical mixing occurs, causing vertical displacement of organisms (Van der Spoel and Schalk 1988), thus creating dispersal possibilities for deeper-living organisms to reach the Indian Ocean. Most (distant) neritic tropical species are distributed in the Indo-Pacific.

Species which occur both in the Red Sea and in the Mediterranean are usually cosmopolitan (Furnestin 1979). *S. neglecta* Aida, 1897 is an example of an Indo-Pacific species which might have reached the Mediterranean through the Suez Canal. It is only recorded in the south-east Mediterranean (Furnestin 1979).

Non-belt-shaped patterns

Oceanic species

Sagitta bierii Alvariño, 1961, in the east tropical Pacific Ocean, has a restricted distribution related to oxygen-poor water (Bieri 1959). Specimens probably belonging to *S.bierii* were found in waters off West Africa, so it is possible that this species is found in both the east Pacific and east Atlantic waters (Ducret 1968; Pierrot-Bults 1974).

S. pseudoserratodentata is another example. It is distributed in the north Pacific central water (Bieri 1959) and was recently recorded by Andreu *et al.* (1989) in the south-west Indian Ocean. *S. serratodentata atlantica* Thomson, 1947 is distributed in the south Pacific transitional water. It is not yet clear whether there is a connection with the Atlantic *S. serratodentata serratodentata* populations through the south Indian Ocean, although a connection is mentioned by Alvariño (1974) (See Fig. 9.13).

S. ferox Doncaster, 1902, *S. robusta* Doncaster, 1902, and *S. regularis* Aida, 1897 are endemic Indo-Pacific species (Bieri 1959; Nair and Rao 1973; Nair 1978). They have no close relatives in the tropical Atlantic (Fig. 9.19).

S. scrippsae Alvariño, 1962 has a north-east Pacific distribution. It is a member of the *S. maxima* group and is most closely related to, or a synonym of, *S. lyra*. The latter species shows a circumglobal 40°N–40°S distribution (Alvariño 1962a, 1969).

Neritic species

Each oceanic coast has, in principle, its own neritic fauna. The patterns can be divided roughly into a warm water neritic pattern (from 50°N–40°S) and a cold water neritic pattern (at latitudes higher than 30°) and show latitudinal zonation. The neritic zonation of the world oceans is shown in Figure 9.20 (Van der Spoel and Heyman 1983). The chaetognath neritic distribution patterns are all warm water patterns.

Neritic species have a more restricted distribution than the oceanic ones. It is to be expected that the separated populations show genetic and taxonomic differences because of the more restricted means of contact between neritic regions. For example, differences have been described between the North Sea and the Oosterscheldt populations of *S. setosa*, the populations described as *S. batava* Biersteker and Van der Spoel, 1966, and also between North Sea and Mediterranean *S. setosa* populations (Furnestin 1979). It is also thought that the Black Sea species *S. euxina* might be conspecific with *S. setosa*, as might *S. bedoti bedoti* Beraneck, 1895 and *S. bedoti minor* Tokioka, 1942, which are distant neritic forms, and *S. bedoti* f. *littoralis* Tokioka and Pathansali, 1965, which is the inlet water form (Tokioka 1979) (Fig. 9.21). The neritic *S. crassa* shows several inlet water forms from different parts of its range, for example, *S. crassa* f. *crassa* Tokioka, 1938, *S. crassa* f. *naikaiensis* Tokioka, 1939 and *S. crassa* f. *tumida* Tokioka, 1939, (Tokioka 1979). It is to be expected that, with more research into infraspecific

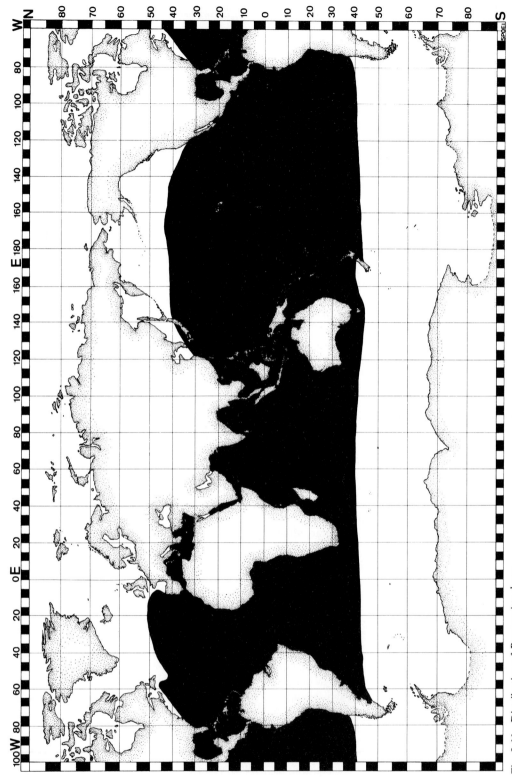

Fig. 9.11 Distribution of *Pterosagitta draco*.

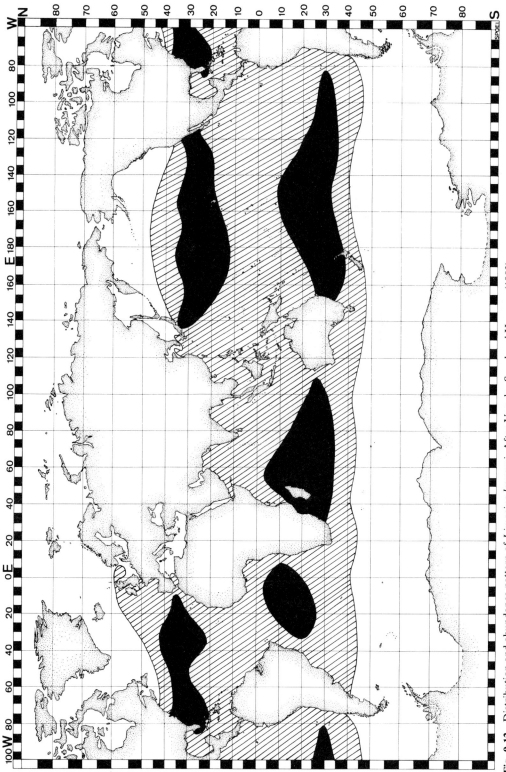

Fig. 9.12 Distribution and abundance pattern of *Limacina lesueuri*. After Van der Spoel and Heyman (1983).

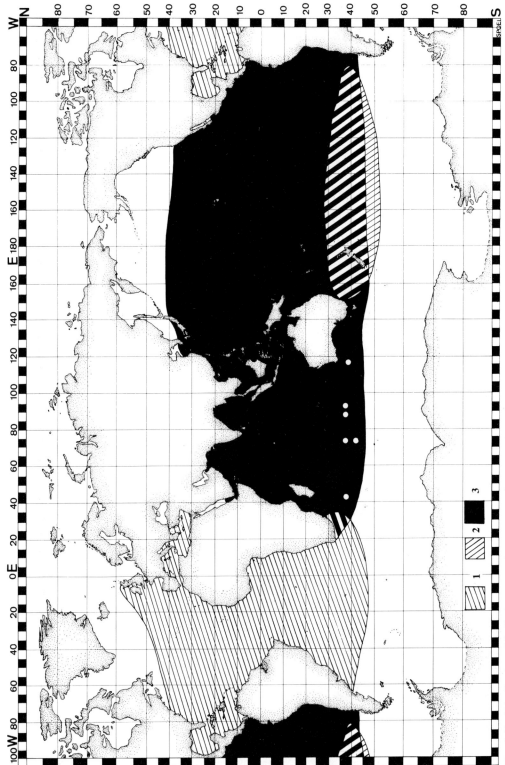

Fig. 9.13 Distribution of 1, *Sagitta serratodentata serratodentata*; 2, *S. serratodentata atlantica*; 3, *S. pacifica*. White dots in the South Indian Ocean represents *S. serratodentata* according to Alvariño (1972).

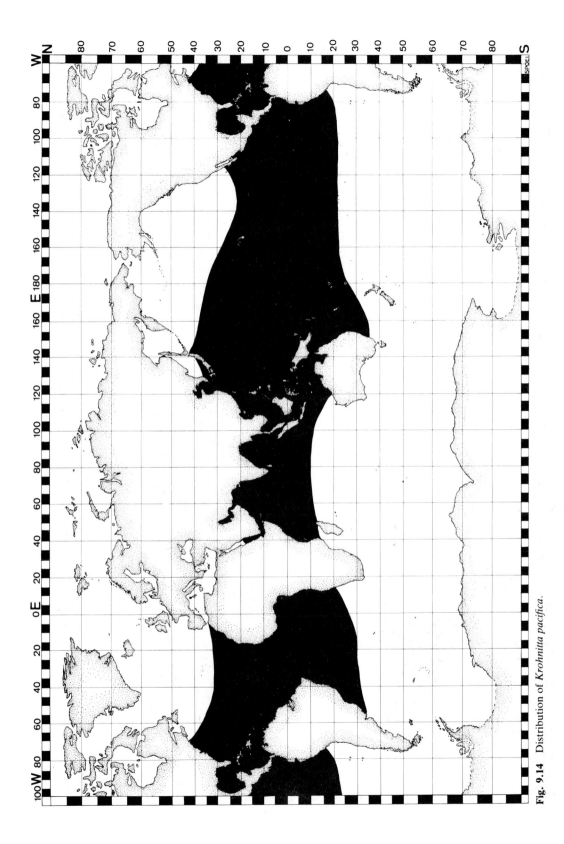

Fig. 9.14 Distribution of *Krohnitta pacifica*.

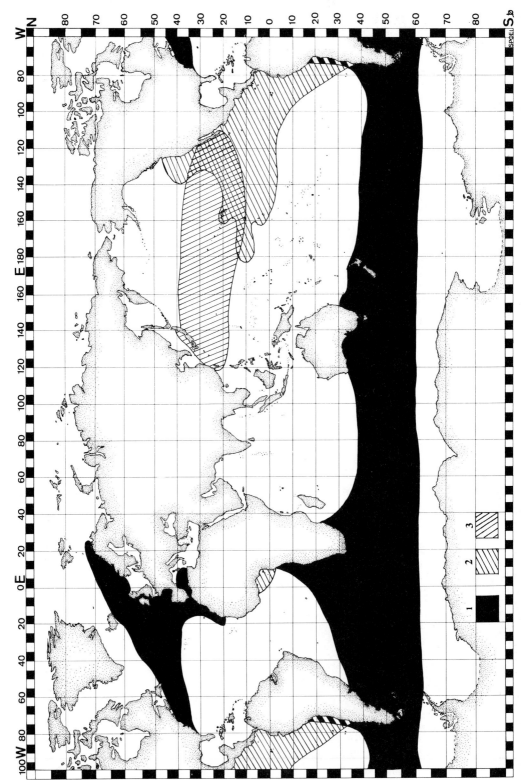

Fig. 9.15 Distribution of 1, *Sagitta tasmanica*; 2, *S. bierii*; 3, *S. pseudoserratodentata*.

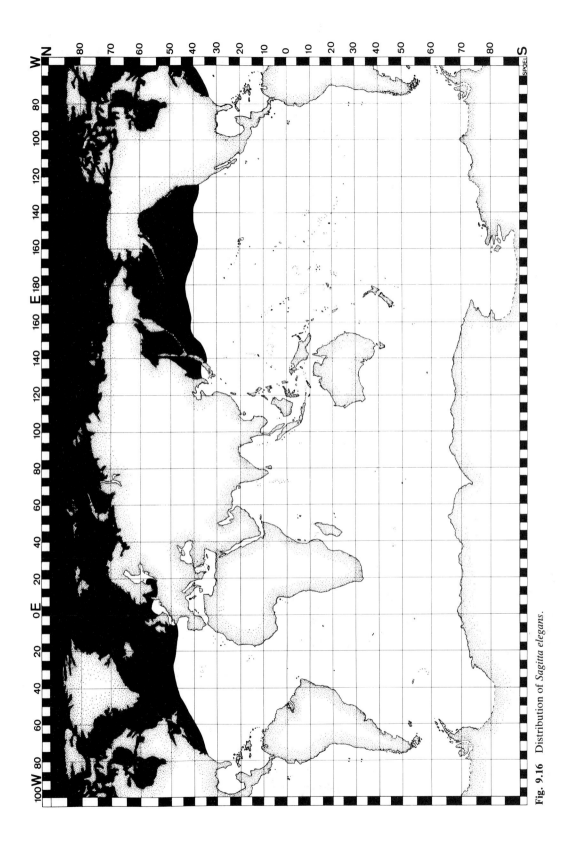

Fig. 9.16 Distribution of *Sagitta elegans*.

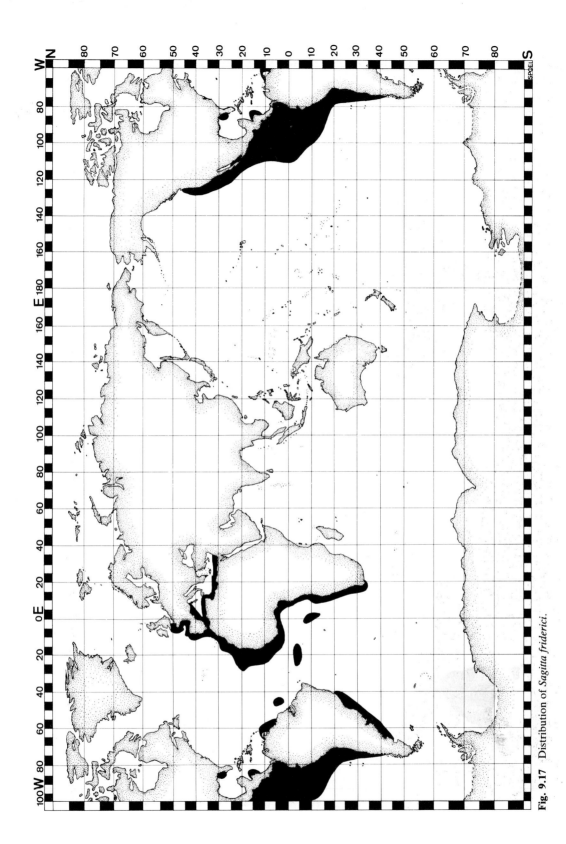

Fig. 9.17 Distribution of *Sagitta friderici*.

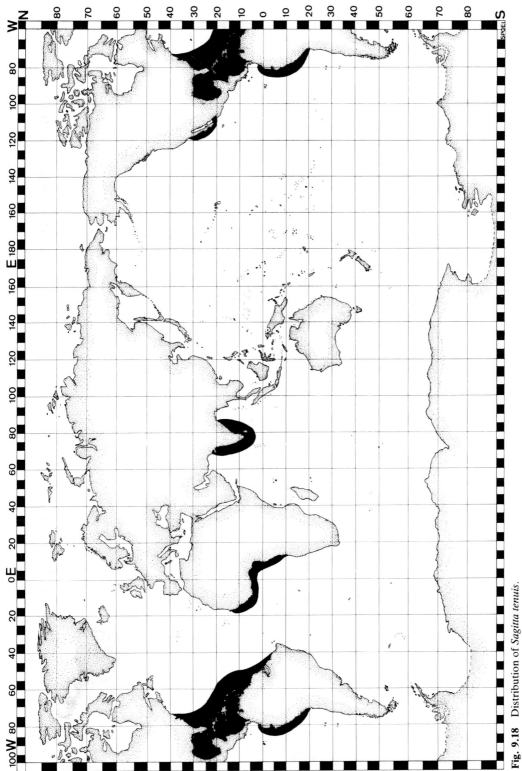

Fig. 9.18 Distribution of *Sagitta tenuis*.

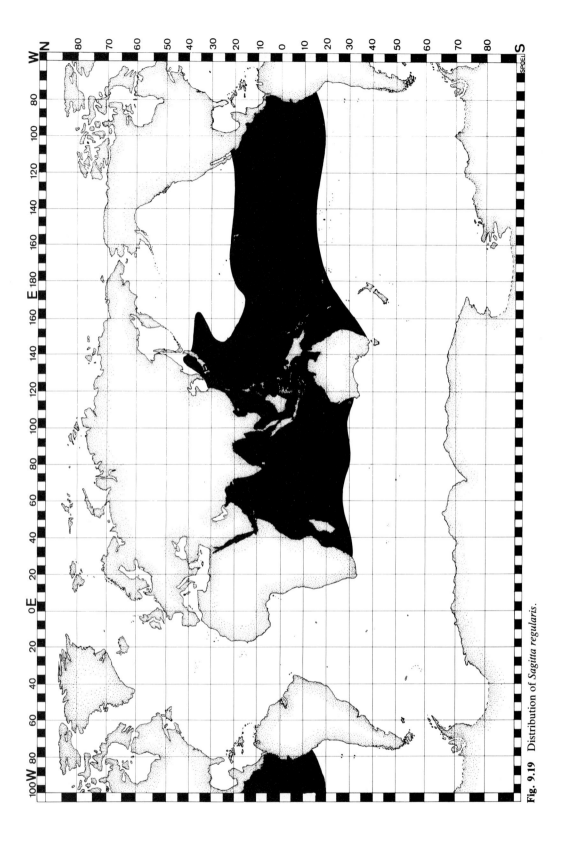

Fig. 9.19 Distribution of *Sagitta regularis*.

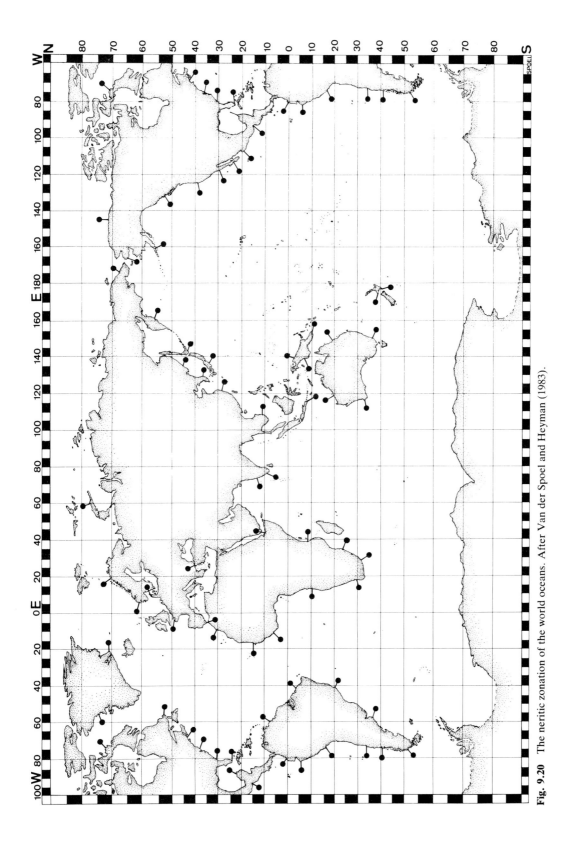

Fig. 9.20 The neritic zonation of the world oceans. After Van der Spoel and Heyman (1983).

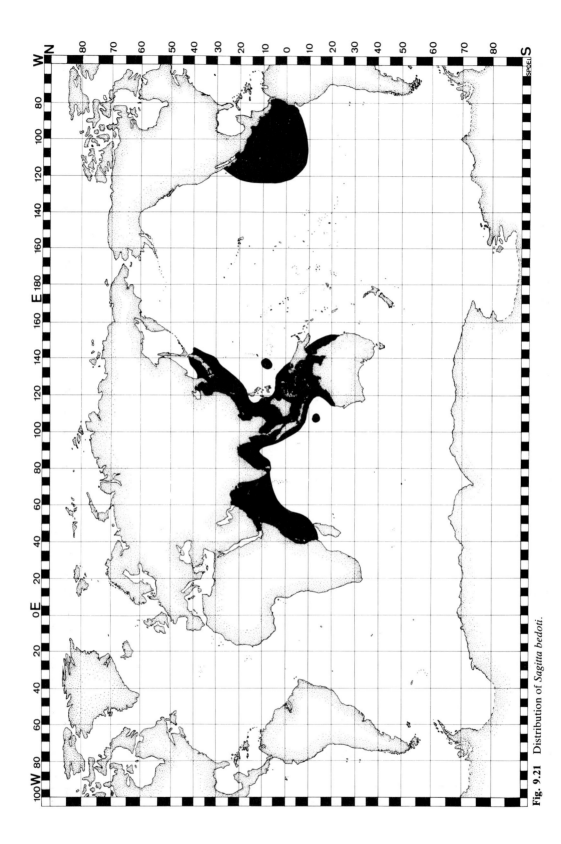

Fig. 9.21 Distribution of *Sagitta bedoti*.

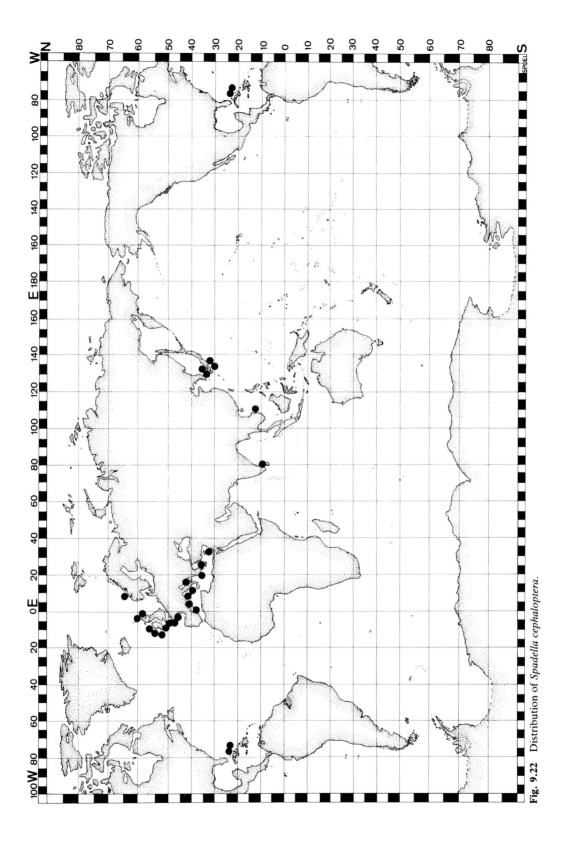

Fig. 9.22 Distribution of *Spadella cephaloptera*.

variation differences between populations in different regions, other neritic species will be found. The Spadellidae are dealt with under benthic patterns.

Warm water patterns

The greatest diversity, especially in neritic species, is seen in the Indo-Malayan region (Tokioka 1959, 1962, 1965*b*; Reid *et al.* 1978). *S. bedoti* is found in the Indian Ocean, the west Pacific, and the east Pacific (Fig. 9.21). The populations on both sides of the Pacific are isolated by distance and it would be interesting to look for infra-specific differences between these populations. The same applies for the west and east Atlantic populations and for those *S. tenuis* populations of the west Atlantic and east Pacific (see Fig. 9.18). The first two have a distance barrier and the last two the Panama isthmus barrier.

Tokioka (1979) discusses neritic taxa from the Indo-West Pacific—*S. bedoti litoralis, S. bedfordii* Doncaster, 1902, *S. crassa, S. crassa* f. *naikaiensis, S. crassa* f. *tumida, S. delicata* Tokioka, 1939, *S. johorensis* Pathansali and Tokioka, 1963 and *S. demipenna* Tokioka and Pathansali, 1963. They all can be considered inlet water forms from bays and lagoons. *S. bombayensis* Lele and Gae, 1936 is an endemic species confined to the near-shore waters of Bombay (Nair 1974).

S. oceania Gray 1930 and *S. tropica* Tokioka, 1942 occur in inlet waters of islands in the Pacific and estuarine zones of India (Nair 1974). *S. bedoti* f. *bedoti* and *S. bedoti* f. *minor* are taxa of the open neritic waters. In general the specimens from inlet waters are much smaller than those from the open neritic waters. This phenomenon is seen in the cold water species *S. elegans* and in the warm water species *S. bedoti.*

S. neglecta is a neritic/distant neritic species of the tropical Indo-Pacific. *S. hispida* Conant, 1896 is distributed in the north and central west Atlantic.

Benthic patterns

The only benthic species which is recorded frequently is *Spadella cephaloptera* Busch, 1851, (Fig. 9.22). There is doubt whether the Atlantic and Pacific populations of *S. cephaloptera* are really conspecific or whether they belong to other species (Bieri, personal communication). The distribution of the other spadellidae is not sufficiently known to speculate about their distribution pattern. The hypothesis is that they will fall into

the neritic patterns following a latitudinal succession, as shown in Figure 9.20.

Rarely recorded species

There have been descriptions of a number of species whose taxonomic position and distribution is not yet clear because of the scarcity of samples. Casanova (1985*a*) described the epipelagic *Sagitta erythraea* from the Red Sea and (1985*b*) the two mesopelagic species *S. lucida* and *S. adenensis* from the north-west Indian Ocean. From the tropical west Atlantic between 0°–8°S Casanova (1986*c*) described *Eukrohnia macroneura* and *E. flaccicoeca* from a depth of more than 300 m. Marumo and Kitou (1966) described the bathypelagic *Heterokrohnia bathybia* from the western North Pacific and Dawson (1968) described *H. involucrum* from the Arctic basin. *H. curvichaeta, H. davidi, H. heterodonta,* and *H. murina* from the north Atlantic have been described by Casanova (1986*a*) and Casanova and Chidgey (1987) described *H. furnestinae*. From the Antarctic *H. fragilis* and *H. longidentata* have been described by Kapp and Hagen (1985) and Hagen and Kapp (1986) described *H. longicaudata*, also from the Antarctic.

Casanova (1986*b*) described a new genus and species, *Archeterokrohnia rubra*, from the North Atlantic and another species *A. palpifera*, from the Mediterranean (Casanova (1986*e*). According to Kapp (1991*a*) *Archeterokrohnia* is synonymous with *Heterokrohnia.*

Species groups

Several species groups are recognized within the chaetognaths, for example the *S. maxima* group, the *S. planctonis* group, and the *S. serratodentata* group.

The *S. planctonis* group consists of *S. planctonis planctonis*, which has a shallow mesopelagic distribution; *S. planctonis zetesios*, which has a deep mesopelagic distribution from 40°N–40°S and an epipelagic/mesopelagic distribution in higher latitudes; and *S. marri*, which is endemic in the Antarctic (see Figs. 9.3 and 9.4), although there is doubt about the taxonomic position of this last species (see Conclusions).

Nearly the same pattern of distribution is shown by the members of the *S. maxima* group (*S. maxima, S. lyra, S. gazellae* and *S. scrippsae*). *S. maxima* is distributed at high latitudes. It shows submergence but a deep tropical connection has not been found, the antarctic populations might be a different taxon (David 1963;

Bieri, personal communciation). *S. lyra* shows a shallow mesopelagic distribution from 40°N–40°S. The endemic Antarctic species is *S. gazellae*. There is an endemic Pacific species described for the north-east Pacific, *S. scrippsae*, but this taxon might be conspecific with *S. lyra*.

The *S. serratodenta* group consists of *S. serratodentata serratodenta* and *S. pacifica*, which had a 40°N–40°S distribution in the Atlantic and Pacific respectively. *S. serratodentata atlantica* has a south Pacific transitional water distribution, *S. pseudoserratodentata* a north

Pacific central water distribution, *S. tasmanica* a north Atlantic cold water and circumglobal subantarctic distribution, and *S. bierii* a tropical east Pacific and east Atlantic distribution (See Fig. 9.15).

The distribution of members of a species group fall within the patterns of restricted distributed species. There is little overlap in distributions within a group. In the opinion of some scientists these species groups are of generic level (Tokioka 1965a; see also Chapter 11).

Conclusions

The same basic distribution patterns (belt-shaped, distant neritic, etc.) are found in different plankton groups, with a minority not distributed like the rest (Johnson and Brinton 1963; Brinton 1962; Fleminger and Hulsemann 1973; Pierrot-Bults 1976a; Van der Spoel and Pierrot-Bults 1979a; Beklemishev 1981; Van der Spoel and Heyman 1983). These distribution patterns are not only reflected in species but also in infraspecific categories (Pierrot-Bults 1976a).

The epipelagic chaetognath fauna of the Atlantic shows little endemism. Only *Sagitta helenae*, *S. hispida* and *S. setosa*, all neritic species, are endemic for the Atlantic. *S. serratodentata serratodentata* is the only oceanic taxon restricted to the Atlantic, while *S. serratodentata atlantica* is confined to the south Pacific (Pierrot-Bults 1974). *S. bombayensis* an endemic species in the Indian Ocean, is also neritic. The Pacific shows the maximum number of endemic taxa, both neritic and oceanic.

Oceanic endemic taxa are *S. scrippsae*, *S. pseudoserratodentata* and *S. bierii*, although there is some doubt about the last two. The distribution of *S. bierii* may also extend to the east-Atlantic (Ducret 1968; Pierrot-Bults 1974) and *S. pseudoserratodentata* is also recorded from the north-west Indian Ocean (Andreu *et al.* 1989) *S. scrippsae* is a north-east Pacific taxon, which is thought to be conspecific with *S. lyra* by some authors (see Chapter 11). Neritic Pacific taxa are *S. delicata*, *S. crassa* f. *crassa*, *S. crassa* f. *naikaiensis*, *S. crassa*, f. *tumida*, *S. bedoti* f. *litoralis*, *S. johorensis* and *S. demipenna*.

Oceanic Indo-Pacific species are *S. robusta*, *S. neglecta*, *S. ferox*, *S. bedoti bedoti*, *S. bedoti minor*, *S. regularis* and *S. pacifica*. *S. pulchra* is a distant neritic Indo-Pacific species and *S. oceania* and *S. tropica* are

neritic species. Only endemics from the Antarctic are known for the epipelagic/mesopelagic taxa: *S. gazellae* and *S. marri*.

Most endemic species are (distant) neritic, while among oceanic species endemic species are subantarctic. This pattern was also found for other planktonic groups (Beklemishev 1981) (See Table 9.3).

The bathypelagic chaetognaths are thought to be widely distributed, although it is not clear if this is true for all bathypelagic species because of the scarcity of samples. *Sagitta macrocephala* is a cosmopolitan deep mesopelagic/bathypelagic species.

Eukrohnia bathyantarctica, *E. bathypelagica* and *Heterokrohnia mirabilis* are species with a circumglobal bathypelagic distribution. The very deep-living, abyssal benthoplanktonic species of *Heterokrohnia*, recently described for the North Atlantic and the Antarctic, might represent provincialism.

Relations of distribution patterns to the relative geological age of the water masses are difficult to ascertain. Following the formation of the different water masses in geological time, only the earliest possible moment for the development of certain taxa can be given.

Van der Spoel and Heyman (1983) put forward a hypothesis for the development of faunas (Fig. 9.23). Related to plate tectonics and the development of the ocean basins, it is postulated that the central water faunas are the oldest possible, while the cold water faunas and the deep sea faunas could only be developed later; cold water taxa are usually ancestral to deep water taxa. Warm water taxa split into equatorial, central water and general warm water taxa. The general warm water taxa can be distinguished as living in the 40°N–40°S belt and the 30°N–30°S belt. It is postu-

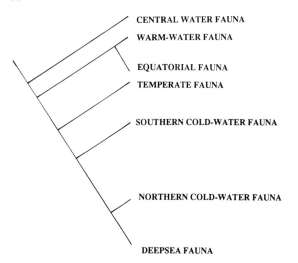

CENTRAL WATER FAUNA

WARM-WATER FAUNA

EQUATORIAL FAUNA

TEMPERATE FAUNA

SOUTHERN COLD-WATER FAUNA

NORTHERN COLD-WATER FAUNA

DEEPSEA FAUNA

Fig. 9.23 Geocladogram of the major planktonic faunas. After Van der Spoel and Heyman (1983).

lated that the 30°N–30°S distributions developed out of the 40°N–40°S distributions in the Eocene–Oligocene, when global temperatures dropped. Some taxa show a limited east Pacific distribution and it is supposed that this pattern is more recent, judging by the taxonomic level of the endeems in this region (Van der Spoel and Heyman 1983).

It is expected that cold water and deep sea fauna elements are more derived than warm and central water elements. The Antarctic endeems are thought to represent the most recent development (Pierrot-Bults and Van der Spoel 1979; Van der Spoel and Heyman 1983). White (1987) postulates that periodic episodes of anoxic events occurring primarily in the mid- to late Cretaceous, Eocene and Miocene caused the developments of the mesopelagic and deep sea faunas.

The splitting of a northern and southern hemisphere cold water fauna is of rather recent origin judging from the close relation of subarctic and subantarctic taxa. Pierrot-Bults and Van der Spoel (1979) discussed the speciation in macrozooplankton. They assumed that the Chaetognatha are oceanic in origin. From the shallow Pacific they dispersed into the Indian and Atlantic Oceans, from tropics to higher latitudes and from the epipelagic to the mesopelagic zone, following the development of the ocean basins and the water masses therein. The bathypelagic realm was the latest to develop, the oldest deep water basin is the North Atlantic. It is usually accepted that the organisms

inhabiting the bathypelagic are derived from cold water ancestors. So the tropical and subtropical plankton of the Pacific is the oldest plankton. Chaetognatha show their greatest diversity in this Ocean (Tokioka 1965b). For chaetognaths only one central water taxon is present (*S. pseudoserratodentata*). General warm water 40°N–40°S distributions are the most common oceanic distribution pattern in Chaetognatha and east Pacific taxa are also present (*S. bieri, S. bedoti*). For Chaetognatha, oceanic equatorial taxa are only present in the Indo-Pacific (*S. robusta, S. regularis*). Salvini-Plawen (1986) stated that (among others) primitive characters in chaetognaths are a long tail (> 40 per cent of the total body length), a great number of teeth, a short fin situated only on the tail and the presence of gut diverticula.

If the Sagittas are divided on the presence or absence of gut diverticula, then the division shown in Table 9.4 is reached.

The elongation of the gut in other groups, such as Crustacea, can be an adaption to poor food conditions (Vinogradov 1970) but since most of the chaetognath species with gut diverticula are neritic or epipelagic, living in areas of high productivity this character is considered not to be adaptive but of phylogenetic importance.

Species with gut diverticula show predominantly epipelagic warm water Indo-Pacific distributions, with the exception of *S. minima*, which shows a circumglobal warm water distribution, and *S. hispida*, a warm water Atlantic species. *S. sibogae, S. decipiens*, and *S. planctonis* are mesopelagic species with gut diverticula. Contrary to the epipelagic species with gut diverticula these show a circumglobal distribution. Species with a circumglobal warm water distribution (except *S. minima*), and Antarctic species are all without gut diverticula. The question arises here whether *S. marri* really belongs to the *S. planctonis* group. This species has no gut diverticulum and the position and shape of the fins is quite different from *S. planctonis planctonis* and *S. planctonis zetesios* (Fig. 9.24). The hypothesis is that the division of *Sagitta* into a group with and a group without gut diverticula took place before the Cretaceous, since the formation of the mesopelagic water masses started in the mid- to late Cretaceous and both groups have mesopelagic taxa.

If Salvini-Plawen's (1986) hypothesis about the primitive nature of the presence of a gut diverticulum, a long tail, and a short fin restricted to the tail segment is correct, then the genus *Sagitta* is more derived than the genus *Pterosagitta*. The latter genus is a monospe-

Table 9.4   Distribution of taxa with and without gut diverticula

	P* warm	I* warm	A* warm	P cold	I cold	A cold	Antarctic
With gut diverticula							
S. johorensis	n*						
S. demipenna	n						
S. delicate	n						
S. bedfordii	n	n					
S. crassa	n	n					
S. neglecta	o*	o					
S. oceania	o	o					
S. regularis	o	o					
S. tropica	o	o					
S. robusta	o	o					
S. ferox	o	o					
S. hispida			n				
S. minima	o	o	o				
S. decipiens				o	o	o	
S. sibogae				o	o	o	
S. pl. planctonis				o	o	o	
S. pl. zetesios				o	o	o	
S. elegans				o		o	
Without gut diverticula							
S. bombayensis		n					
S. pseudoserratodentata	o						
S. scrippsae	o						
S. pulchra	n	n					
S. pacifica	o	o					
S. bedoti	n	n	n				
S. tenuis	n	n	n				
S. friderici	n	n?	n				
S. bipunctata	o	o	o				
S. bierii	o	o	o				
S. enflata	o	o	o				
S. s. atlantica	o	o	o				
S. s. serratodentata			o				
S. helenae			n				
S. setosa			n				
S. hexaptera				o	o	o	
S. lyra				o	o	o	
S. maxima				o	o	o	
S. macrocephala				o	o	o	
S. tasmanica northern form						o	
S. tasmanica southern form							o
S. gazellae							o
S. marri							o

* I, = Indian Ocean; P = Pacific Ocean; A, = Atlantic Ocean; n, = neritic; o, = oceanic.

cific genus distributed in the tropical/subtropical epipelagic, so the distribution pattern also points to a
primitive genus. (See Fig. 9.11).

Another hypothesis following from the distribution
patterns is that the mesopelagic and/or bathypelagic
genera *Krohnitta, Eukrohnia*, and *Heterokrohnia* are
more derived than both *Pterosagitta*, which is epipelagic, and *Sagitta*, which is epipelagic, mesopelagic and
bathypelagic.

Fig. 9.24 *Sagitta marri*. After Pierrot-Bults (1975).

10

DEEP SEA CHAETOGNATHS
Makato Terazaki

Introduction

All depths of the oceans below 1000 m are considered here as deep sea. There are relatively few samples from this region, and even fewer from more precisely defined depths, so that knowledge of deep sea chaetognaths is incomplete and fragmentary and only single individuals or a few specimens have been caught so far for a number of species. The distribution and way of life of many of the deep sea species is thus very incompletely known. As several new species have been found recently, many more presumably remain to be collected and described. The genus *Heterokrohnia* is bathypelagic (Kapp and Hagen 1985; Casanova 1986*a,b*), as are *Eukrohnia macroneura and E. flaccicoeca* (Casanova 1986*c*). *Heterokrohnia* is here taken to include two *Archeterokrohnia* species (Kapp 1991*a*) and the recently described *H. mirabiloides* (Casanova and Chidgey 1990) is held to be synonymous with *H. mirabilis* (Kapp 1991*b*) hence the genus presently contains 13 species. *Krohnitella* (two species) and the monotypic genera *Bathybelos* and *Bathyspadella* are bathypelagic or deep benthopelagic. Some *Eukrohnia* and *Sagitta* species are bathypelagic or temporarily bathypelagic, living at great depth in tropical regions (e.g. *E. hamata*) or sinking to great depths during development, being bathypelagic when mature (e.g. *S. gazellae* and *S. marri*). Others are mesopelagic and bathypelagic (e.g. *S. macrocephala* and *S. zetesios*).

Morphology

Generally, deep sea chaetognaths are structurally more robust than epipelagic species. They also have many large hooks to hold their prey (Nagasawa and Marumo 1978; Terazaki, unpublished data). The most striking morphological characteristics of deep sea species are the different head armatures in the *Heterokrohnia* species; teeth of different length, form, number and arrangement; and various forms of vestibular organs (Kapp and Hagen 1985; Casanova 1986*a, b*). *Krohnitella* and *Bathyspadella* are exceptional in lacking teeth and *E. macroneura* and *E. flaccicoeca* have fewer teeth and more hooks than *E. hamata* and *E. bathypelagica*. Tokioka (1950) indicated that the eyes are relatively large in bathypelagic species, whereas the pigment cells of the eyes are larger in epipelagic species. Eye structure in relation to depth is considered in Chapter 3. *Heterokrohnia* species are blind, lacking eyes (Marumo and Kitou 1966; Kapp and Hagen 1985; Casanova 1986*a*; Hagen and Kapp 1986), as are two recently described *Eukrohnia* species (Casanova 1986*c*).

Reproduction

Terazaki and Miller (1986) found that *E. hamata* has three spawning periods during the year in the Gulf of Alaska. These periods are separated by distinct gaps both in the presence of mature individuals and the appearance of new juveniles; in *E. bathypelagica* and *E. fowleri*, breeding is continuous. Breeding periodicity seems to be temperature-dependent (Dunbar 1962; Sameoto 1971, 1973; and see Chapter 7). Deep-living *Eukrohnia* species have developed breeding or marsupial sacs, presumably to protect their developing offspring in a harsh environment (Alvariño 1968; Dawson 1968; Terazaki and Miller 1982). Such sacs have been found in *E. hamata* (Ritter-Záhony 1910, 1911; Dawson 1968); *E. fowleri* (Ritter-Záhony 1910, 1911; Terazaki and Miller 1982); *E. bathyantarctica* (Alvariño 1968); and *E. bathypelagica* (Terazaki and Miller 1982).

In *E. hamata*, two breeding sacs, each containing approximately 50 young, are developed on each side. All but the hindermost of the young were orientated with their anterior end towards the anterior end of the sac (Dawson 1968). In *E. bathypelagica*, the fertilized egg is some 480 μm in diameter, and the number in both marsupia ranged from 19–30 (Fig. 10.1(A)). The gelatinous marsupium, covered by the lateral fins, is widely attached near the tail septum. The fertilized eggs (Fig. 10.1(B)) develop into young in the sac (Fig. 10.1(C)). No mature eggs were found in the ovaries of specimens with marsupial sacs, hence it seems that all fertilized eggs are released through the oviduct at once. In *E. fowleri*, the fertilized eggs are larger, 900 μm in diameter and five to six eggs are found in each sac. The sacs hang from a trumpet-shaped structure at the oviducal openings (Fig. 10.1(D)). The sacs themselves are ovoid, 1.1–1.3 mm × 2.0–2.5 mm, and differ from those of *E. bathypelagica* in having a harder surface membrane and containing more tightly-packed eggs.

Chaetognaths generally die after spawning (McLaren 1969), but many individuals of these two deep-living *Eukrohnia* species were collected in the Gulf of Alaska carrying broken marsupial sacs; some had new ova in the ovaries. This supports the view that these species have a long breeding season and make several releases of fertilized eggs (Terazaki and Miller 1982).

Epipelagic species release quite a large number of eggs at each spawning (Murakami 1959; Dallot 1968; Reeve 1970a). In contrast *E. bathypelagica* and *E. fowleri* release very few eggs but the young develop in

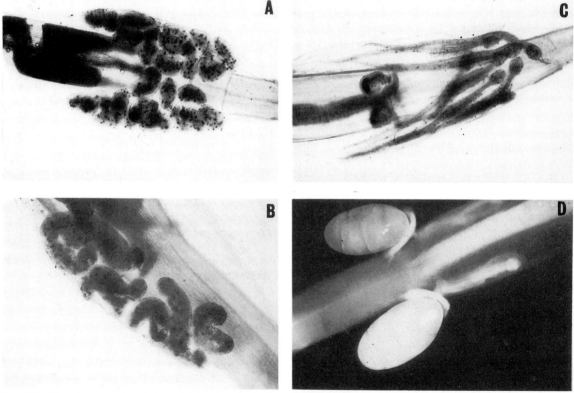

Fig. 10.1 Posterior part, dorsal view of *Eukrohnia bathypelagica* carrying the fertilized eggs (A, B) and the young (C) in the marsupium. (D) Posterior part, dorsal view of *Eukrohnia fowleri* carrying the fertilized eggs in marsupial sacs. Reproduced, with permission of the publisher, from Terazaki and Miller (1982).

a marsupium (this is in line with the general strategy of lower fecundity as care of the young increases). The more developed young of *E. hamata* are quite uniform in size (2.4 mm long, Dawson 1968), whilst those of *E. bathypelagica* are 2.5 mm long at hatching, and are covered with alveolar tissue (see Chapter 2); three to six pairs of hooks are present on the head. Newly-hatched *E. fowleri* are 3.0–3.5 mm long, and their whole body is also covered with massive alveolar tissue; the body structure is very strong compared to that of *E. bathypelagica*. Newly-hatched young of epipelagic chaetognaths are much smaller (1.0 mm for *Sagitta bipunctata* (Doncaster 1902); 1.2–1.4 mm for *S. elegans* (Kotori 1975*a*); 0.5–0.6 mm for *S. nagae* (Nagasawa and Marumo 1978); and 0.7 mm for *S. crassa* (Murakami 1959)). Presumably the larger size of the young of deep sea chaetognaths is related to the develop-

ment of the marsupia and to survival in the deep sea.

Casanova (1985*c*) observed an annex gland (presumably secretory) in some species of *Heterokrohnia*. This surrounds the gut and opens into the base of the ovary. Casanova (1985*c*) also described a duct common to both annexe gland and ovary, connecting them to the corresponding cavity of the tail.

Chaetognaths are hermaphroditic, hence both cross-fertilization by mutual mating and self-fertilization are, in principle, possible. Since bathypelagic chaetognaths are at low density in the deep sea, and the opportunity for mating is thus rare, self-fertilization has been suggested. However, the finding of a specimen of *E. fowleri* with full seminal vesicles on the tail and fertilized eggs in the ovary (Terazaki and Miller 1982) suggests that *E. fowleri* (and other deep sea species) depend upon cross-fertilization.

Feeding

Deep sea chaetognaths may be expected to feed independently from any day–night rhythm, as they presumably rarely have the opportunity to capture prey in the sparsely populated deep sea environment. In comparison with epipelagic chaetognaths (Chapter 5), not much is known about the feeding habits of mesopelagic and bathypelagic chaetognaths because sampling is laborious and they are often damaged during net sampling from the deep sea. However, the feeding habits of the deep mesopelagic species *Sagitta zetesios* have been examined by Terazaki and Marumo (1982). The predominant food organisms were copepods; the percentages of Copepoda, Chaetognatha, Euphausiacea, Ostracoda and unidentified in the gut were 71.6 per cent, 16.2 per cent, 1.4 per cent, 1.4 per cent and 9.4 per cent, respectively. Twenty-four copepod species, five chaetognath species and one ostracod were identified, and of these, 20 species were mesopelagic

and bathypelagic; only 10 were epipelagic (Table 10.1). Young *S. zetesios* living between 300–500 m were very different from adults living below 500 m. Young specimens fed mainly during the night, preferring small copepods and chaetognaths, and their FRC (number of specimens containing food in the gut/total number of specimens examined) ranged from 16.7 to 22.6. Adults preferred larger food organisms (mainly copepods 2–3 mm in size) and fed more actively by day. Their FRC was under 8.7.

The epipelagic *S. elegans* also inhabits the bathypelagic layer of the Sea of Japan. Their FRC was near 0 in the bathypelagic layer, although it was more than 20 in the epipelagic layer (Terazaki unpublished data). Finally, according to Casanova (1986*a*), *Heterokrohnia* and *Archeterokrohnia* (see Kapp 1991*a*) feed upon sediment bacteria, although this remains to be confirmed.

Pigmentation

Most pelagic chaetognaths are colourless and transparent or milky white, although Bieri (1966, 1977) reported pale blue and deep blue epipelagic chaetognaths. However, some deep sea species have yellow, red or orange pigments, for example, *Sagitta macrocephala* (Tokioka 1939), *Eukrohnia fowleri* (Owre 1960), *Heterokrohnia*

bathybia (Marumo and Kitou 1966), *H. longicaudata* (Hagen and Kapp 1986), and *H. (Archeterokrohnia) rubra* (Casanova 1986*b*). The pigment generally occurred in the gut (and was restricted to the oseophagus in *H. longicaudata*), but occasionally extended to the head and fins in *E. fowleri*. The gut of *S. macro-*

Table 10.1 List of the food organisms found in the gut of *Sagitta zetesios* collected from Sagami Bay, Central Japan

Food organisms	Juvenile	Stage I	Stage II	Stage III	Stage IV
Copepoda			+		
Calanus cristatus					
C. plumchrus	+	+			
C. pacificus	+	+	+		+
* *Undinula vulgaris*	+				
Eucalanus bungii	+				
* *E. crassus*		+			
* *E. subtenuis*		+		+	
Rhincalanus nasutus		+	+		
* *R. cornutus*			+		
* *Clausocalanus arcuicornis*	+				
Aetideus armatus		+			
Gaetanus armiger			+		
* *Euchaeta concinna*		+	+		
Pareuchaeta russelli		+	+		+
P. scaphula			+		
P. rubra				+	
Scaphocalanus echinatus			+		
* *Scolecithrix danae*		+			
* *S. nicobarica*		+			
Scolecithricella valida			+		
Pleuromamma abdominalis			+		
Heterorhabdus pacificus		+			
* *Candacia bipinnata*		+			
Oncaea conifera		+			
Chaetognatha		+			
* *Sagitta nagae*					
S. lyra		+			
S. neodicipiens			+		
S. macrocephala			+		
S. zetesios			+	+	
Ostracoda					
Conchoecia elegans		+			

* Epipelagic species. After Terazaki and Marumo (1982)

cephala and *E. fowleri* was found to contain a carotenoid pigment with many small oil droplets. Tokioka (1939) and others (Katayama *et al.* 1965; Lee 1966) supposed that the carotenoids of the prey were transferred to the predator body without any chemical change. However, Terazaki *et al.* (1977) studied the pigments chromatographically and found that the characteristics of the carotenoids in chaetognaths were different from those of dietary pigments (Table 10.2).

There was no difference between the pigmentation of the intestine of the mature and immature specimens; all pigments were carotenoids. Like all carotenes, the major carotenoid was very soluble in polar solvents, although its absorption spectrum formed a single broad peak around 460 nm. Thus it is inferred that carotenoids in chaetognaths are synthesized by the chaetognaths themselves, rather than being dietary pigments deposited in the intestinal tissue.

Table 10.2 Comparison of major carotenoids in deep sea chaetognaths and copepods

Species	Absorption maxima (nm)				R_f value (1% acetone in petroleum ether)
	Petroleum ether	Hexane	Benzine	Chloroform	
Chaetognaths					
Sagitta macrocephala	457–460	460	475–477	478–480	1.0
Eukrohnia fowleri	457–460	459–460	473	478–480	1.0
Copepods					
Calanus pacificus	465	465	ND	484	0.4
Lucicutia bicornuta	468	465	485	483	0.3

ND, Not determined. Reproduced, with permission of the publishers, from Terazaki *et al.* (1977)

11

SYSTEMATICS OF THE CHAETOGNATHA
Robert Bieri

In recent years the number of species, genera, and families within the Chaetognaths has exploded. In his revision of the phylum Ritter-Záhony (1911*a*) recognized 27 species in six genera. In his 1965*a* revision, Tokioka recognized 58 species in 15 genera. Twenty-three years after Tokioka's revision, 115 species in 23 genera are currently under discussion among chaetognath specialists and this latter list is conservative—it omits an additional 31 species and five genera discussed by Kassatkina (1982). If these were included, the number of genera would jump to 28 and the number of species to 146.

We can expect the number of species and genera to increase because many new species remain to be discovered, especially among the benthic and bathyal families. The greatest increases in described species have occurred in the spadellids, eukrohnids and the heterokrohnids. The number of valid species may ultimately reach more than 200.

The benthic spadellids are clearly vastly under-reported. Only two species—both previously reported from the Pacific—have been reported from the Indian Ocean and only four species from the entire southern hemisphere (Johnson and Taylor 1920; Mawson 1944). Also under-reported is the Pacific Ocean, with two species from Japan, three species from California, and two species from the mid-Pacific. With four or five species reported from the western North Atlantic and five species from the eastern North Atlantic, it seems possible that this one family will produce a minimum of twenty more species.

The recent burgeoning of the genus *Heterokrohnia* from one recognized species in 1950 to ten species today indicates that there are other species still undescribed. If the shallow water benthic family Spadellidae is vastly under-reported, what portends for the deep water benthic or hyperbenthic family Krohnittellidae (Bieri 1974*a*)?

The Chaetognatha now show a much greater morphological diversity than previously recognized. *Bathybelos*, with a dorsal nerve ganglion instead of the usual ventral nerve ganglion (Owre 1973; Bieri 1991), is perhaps the most striking new discovery, but the tubules connecting the ovaries and testes in *Heterokrohnia* and *Archeterokrohnia* are also a radical departure from the usual chaetognath body plan (Casanova 1985*c*). The three sets of teeth, with their vastly different structure and arrangement, reported by Zhang and Chen (1983) in *Eukrohnia*(?) *sinica* are an amazing new development. Differences in gut structure (Dallot 1970), and in the strange seminal receptacles of *Pterokrohnia arabica* (Srinivasan 1986) also point to the diversity within the phylum. With the advent of scanning electron microscopy several new types of microstructures have been demonstrated, but detailed comparative studies remain to be completed. For example, the narial pores in *Flaccisagitta hexaptera* have not yet been demonstrated in any other species (Thuesen and Bieri 1987). Some morphological differences have been observed in the transvestibular pores (Thuesen *et al.* 1988*b*). They have been reported from 18 sagittid species but are not found in *Eukrohnia*. The microstructure of the eyes is extremely diverse (Ducret 1977, 1978; Goto *et al.* 1988; see also Chapter 3). This diversity of structure is only partially integrated into the present system of classification. It seems inevitable that further changes will have to be made in the delineation of genera and families as new species are discovered and the microstructure of extant species elucidated further.

Meanwhile, the gross morphology of some of the better established species needs more careful and complete analysis. For example, *Caecosagitta macrocephala* is still very poorly known in terms of the mature seminal vesicle, mature ovaries, the extent of rays in the fins and in terms of general body proportions (compare the illustrations of this species in Tokioka (1940), Alvariño (1967), Michel (1984), and McLelland (1989*b*)). Casanova (personal communication) has expressed the need to know more exactly the extent of the ventral transverse musculature in the spadellids (structures which have not been adequately described in the past) and McLelland (personal communication) has pointed out that we need to know more of the structure of the intestinal diverticula; these vary in different species (Dallot 1970; see Table 9.4).

Not only do we have increased awareness of greater morphological variation within the phylum, but we are also beginning to recognize a much greater diversity in habitat utilization by different groups of chaetognaths. The simple divisions of benthic and planktonic, although still fundamental, are no longer entirely adequate. In Table 11.1 some of these differences are

Table 11.1 Conservative* estimate of species currently recognized or under discussion in the phylum Chaetognatha

	Number of species
Benthic genera	
Spadella	8
Paraspadella	9–10
Bathyspadella	1
Krohnittella	2
Planktonic genera	
Archeterokrohnia (hyperbenthic?)	3
Eukrohnia (partly hyperbenthic)	10
Heterokrohnia (partly hyperbenthic)	10–11
Bathybelos	1
Krohnitta	2–3
Pterokrohnia	1
Pterosagitta	1
(*Sagitta* sensu Ritter-Záhony, 1911)	52–62
*Sagitta***	3
Aidanosagitta	14
Caecosagitta	1
Ferosagitta	5–7
*Flaccisagitta***	3
Mesosagitta	4–5
*Parasagitta***	5–8
*Pseudosagitta***	3–4
Serratosagitta	5–6
Solidosagitta	4
Zonosagitta	5–7
Fossil genera	
Paucijaculum and related spp.	1–3
Total number of species:	100–115

* Many species and genera of Kassatkina omitted because status uncertain; ** usage of Bieri herein.

indicated in the 'hyperbenthic' nature of *Archeterokrohnia*, *Heterokrohnia* (Casanova 1986a,b; Casanova and Chidgey 1987) and one of the eukrohnids (McLelland 1989a). The benthic spadellids are sometimes taken in plankton tows. See for example Conant (1896) and the new species of blind cave-dwelling paraspadellid recently described by Bowman and Bieri (1989) which was taken by a diver in mid-water. An apparently new species of *Aidanosagitta* (Bieri and Bowman, in press) from the Seychelle Islands was taken in a meiobenthic sample. Some, if not most, species in this genus may be analogous to the hyperneritic pontellids of the genus *Labidocera* studied by Fleminger (1979) and Fleminger *et al.* (1982). *Parasagitta hispida*, an abundant species off Florida, appears to be quasiplanktonic because, although it shows no particular modification in gross morphology, in aquaria it often attaches to the glass sides of the tanks (Michel, personal communication). Thus, chaetognaths can be benthic (*Paraspadella*), epibenthic (some *Spadella*), hyperbenthic (*Archeterokrohnia*, some *Heterokrohnia*, and some *Eukrohnia*), planktonic, quasiplanktonic (*Parasagitta hispida*), and hyperneritic (some *Aidanosagitta*). I propose the name hyperneritic as a more general term than Tokioka's 'Inlet Species' although inlets may prove to be critical to the maintenance and evolution of these species. Special types of equipment are needed to take adequate samples from these significantly different habitats. Species which seem rare may prove to be very abundant when proper collecting apparatus is used.

The fossil *Paucijaculum samamithion*, reported by Schram (1973) from the Mazon Creek assemblage in the Upper Pennsylvanian of Illinois is probably a chaetognath. Although not described by Schram, Bieri (unpublished data) found traces of mature ovaries, hooks, possible seminal vesicles and possible alveolar tissue, enough different traces in enough different specimens to convince him that, surprisingly, these are indeed valid traces of chaetognaths. This stands in marked distinction to Walcott's presumed fossil chaetognath, *Amiskwia* from the Cambrian Burgess Shale. This fossil has been disclaimed as a chaetognath by Owre and Bayer (1962), Conway-Morris (1977), and by Bieri (unpublished data). *Amiskwia* has no substantive structures that mark it as a chaetognath— no hooks, no ovaries, no seminal vesicles. This is surprising, because traces of soft tissues are generally much better preserved and are much easier to observe in the Burgess Shale than in the Mazon Creek fossils. As well as lacking critical chaetognath structures,

Amiskwia has definite structures that are unknown in any extant chaetognath—a gut that extends to the tip of the tail and two extremely large 'tentacles'.

The Mazon Creek material appears to contain three species in three genera that are closely related to extant planktonic genera, not to the benthic genera (Bieri unpublished). The close relationship to extant genera argues for an origin and differentiation much earlier than the Carboniferous, possibly in the Cambrian explosion.

For the convenience not only of experienced workers but also for those unfamiliar with or newly interested in the phylum, a list of families, genera, and species is given in Table 11.2. Simple keys to the families (Table 11.3) and to the genera in the family Sagittidae (Tables 11.4 and 11.5) are also given. In Table 11.2 the type species of each genus is listed first, the remaining species are listed in alphabetical order. A few questions

of synonomy are indicated but by no means all. No phylogenetic assumptions are implied by the order of the families which, for ease of comparison, approximately follows Tokioka (1965*a*).

Illustrations of representative species of each genus are included in Table 11.2 with the usual mature size in mm. Also of use are Tokioka's illustrations (1940, 1942) and his figures reproduced by Yamaji (1980); Alvariño's figures in her 1963 and 1967 publications; Michel's (1984) paper; Pathansali (1974) (for the genus *Aidanosagitta*); and finally McLelland (1989*b*). Because of the confused state of chaetognath systematics at all levels, from species to classes, and because new information on the fossil record and molecular relationships may become available in the next two or three years, no phylogenetic speculations are included in this discussion.

General functions of systematics applied to the Chaetognatha

At present the state of chaetognath systematics is close to chaotic. For this reason some general statement of systematics as applied to the Chaetognatha seems in order. As an example of confusion, if not chaos, see Alvariño (1967), Pathansali (1974), Tokioka (1974*b*, 1979), and earlier papers for the confusion of the genus *Zonosagitta*. Another example is the incorrect switching by Alvariño (1962*b*) of the trivial names *ferox* and *robusta* in the genus *Ferosagitta*. Many authors, including Alvariño, seem to be unaware of Thomson's (1947) clear discussion of the types of *robusta* and *ferox*. The matter is also discussed by Tokioka (1965*a*) and by Pathansali (1974). For the next few years it will be necessary to qualify these trivial names with 'big species', 'little species', or 'non-Alvariño', as done here. Disagreement not only at the species level but also at the generic level, has produced much confusion not only among systematists but also among behaviourists, comparative morphologists and ecologists.

The continual changing of names is confusing and can slow or greatly impede progress in science. On the other hand, the prolonged use of incorrect names and ideas also impedes science and may equally result in confusion and frustration. Tokioka's revision, incorrect as it may be in several respects, has not been accepted by all chaetognath workers, and I regard this as an example of conservatism that has greatly hindered progress in every field. For example, in an otherwise outstanding recent paper, Goto *et al.* (1989) gave a

detailed analysis of the variation of eye structure as a function of depth of habitat. However, they included all eleven species they examined in Ritter-Záhony's genus *Sagitta*. The diversity of structure they demonstrated is truly amazing and is better understood if the analysis follows the more recent generic divisions. A study of their data using the systematic arrangement in Table 11.5 shows striking intergeneric differences in the microstructure of the eyes (Bieri, unpublished data).

While taxonomists should strive for simplicity and convenience, the basic functions of any taxonomy are:

1. To establish names of discrete units so that communication can occur.

2. To create categories (names) with significant informational content.

3. To define relatively homogeneous groupings from which generalizations and predictions can be made (these predictions based in turn on appropriate samples).

4. In the case of biological systems, the classification may eventually demonstrate evolutionary relationships, but this function should follow from the three primary objectives of naming discrete categories, including maximal information content,

Table 11.2 List of chaetognath species by family and genus. (The type species is listed first, the remaining species are in alphabetical order. Representative illustrations are mostly of the type species. The usual length at maturity (mm ± 30%) is given next to each figure.)

SPADELLIDAE Tokioka, 1965

Genus *Spadella* Langerhans, 1880
Spadella cephaloptera (Busch, 1851) (Fig. 11.1)
Spadella angulata Tokioka, 1951
Spadella birostrata Casanova, 1987
Spadella bradshawi Bieri, 1974
Spadella equidentata Casanova, 1987
Spadella gaetanoi Alvariño, 1978
Spadella ledoyeri Casanova, 1986
Spadella moretonensis Johnston and Taylor, 1920

Fig. 11.1
8 mm

Genus *Paraspadella* Salvini-Plawen, 1987 (see Bowman and Bieri 1989)
Paraspadella schizoptera (Conant, 1895) (Fig. 11.2)
Paraspadella anops Bowman and Bieri, 1989
Paraspadella caecafea (Salvini-Plawen, 1987)
Paraspadella gotoi (Casanova, 1990)
Paraspadella hummelincki (Alvariño, 1970). Syn. of *P. pulchella*?
  (Owre 1973)
Paraspadella johnstoni (Mawson, 1944)
Paraspadella legazpichessi (Alvariño, 1981)
Paraspadella nana (Owre, 1963)
Paraspadella pimukatharos (Alvariño, 1987)
Paraspadella pulchella (Owre, 1963)
Paraspadella sheardi (Mawson, 1944)

Fig. 11.2
8 mm

Genus *Bathyspadella* Tokioka, 1939
Bathyspadella edentata Tokioka, 1939 (Fig. 11.3)

EUKROHNIIDAE Tokioka, 1965

Genus *Eukrohnia* Ritter-Záhony, 1909
Eukrohnia hamata (Möbius, 1875) (Fig. 11.5)
Eukrohnia bathyantarctica David, 1958
Eukrohnia bathypelagica Alvariño, 1962
Eukrohnia calliops McLelland, 1989
Eukrohnia flaccicoeca Casanova, 1986
Eukrohnia fowleri Ritter-Záhony, 1909
Eukrohnia kitoui Kuroda, 1981
Eukrohnia macroneura Casanova, 1986
Eukrohnia minuta Silas and Srinivasan, 1968
Eukrohnia proboscidea Furnestin and Ducret, 1965
Eukrohnia(?) *sinica* Zhang and Chen, 1983

Fig. 11.3
12 mm

Fig. 11.4
23 mm

Table 11.2   (*continued*)

Genus *Archeterokrohnia* Casanova (1986)
Archeterokrohnia rubra Casanova, 1986 (Fig. 11.4)
Archeterokrohnia longicaudata (Hagen and Kapp, 1986)
Archeterokrohnia palpifera (Casanova, 1986)

Genus *Heterokrohnia* Ritter-Záhony, 1911
Heterokrohnia mirabilis Ritter-Záhony, 1911 (Fig. 11.6)
Heterokrohnia bathybia Marumo and Kitou, 1966
Heterokrohnia curvichaeta Casanova, 1986
Heterokrohnia davidi Casanova, 1985
Heterokrohnia fragilis Kapp and Hagen, 1985
Heterokrohnia furnestinae Casanova and Chidgey, 1987
Heterokrohnia heterodonta Casanova, 1986
Heterokrohnia involucrum Dawson, 1968
Heterokrohnia longidentata Kapp and Hagen, 1985
Heterokrohnia murina Casanova, 1986

Krohnittellidae Bieri, 1989

Krohnittella boureei Germain and Joubin, 1912 (Fig. 11.7)
Krohnittella tokiokai Bieri, 1974

Krohnittidae Tokioka, 1965

Krohnitta subtilis (Grassi, 1881) (Fig. 11.8)
Krohnitta pacifica (Aida, 1897)
Krohnitta mutabbii Alvariño, 1969. Syn. of *K. pacifica*? (Michel 1984)

Bathybelosidae Bieri, 1989

Bathybelos typhlops Owre, 1973 (Fig. 11.9)

Sagittidae Claus and Groben, 1905

Genus *Sagitta* Quoy and Gaimard, 1827
Sagitta bipunctata Quoy and Gaimard, 1828 (Fig. 11.10)
Sagitta bombayensis Lele and Gae, 1936
Sagitta helenae Ritter-Záhony, 1911
See also *Ferosagitta* and *Parasagitta*

Genus *Aidanosagitta* Tokioka and Pathansali, 1963
Aidanosagitta neglecta (Aida, 1897) (Fig. 11.11)
Aidanosagitta alvarinoae (Pathansali, 1974)
Aidanosagitta bedfordii (Doncaster, 1902)
Aidanosagitta crassa (Tokioka, 1938)
Aidanosagitta delicata (Tokioka, 1939)

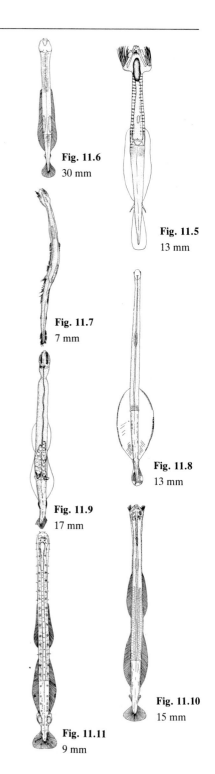

Fig. 11.6
30 mm

Fig. 11.5
13 mm

Fig. 11.7
7 mm

Fig. 11.8
13 mm

Fig. 11.9
17 mm

Fig. 11.10
15 mm

Fig. 11.11
9 mm

Table 11.2 (*continued*)

Aidanosagitta demipenna (Tokioka and Pathansali, 1963)

Aidanosagitta erythraea (Casanova, 1985)

Aidanosagitta guileri (Taw, 1974)

Aidanosagitta johorensis (Pathansali and Tokioka, 1963)

Aidanosagitta oceania (Grey, 1930) (Trivial name is often mispelled
  oceanica.)

Aidanosagitta ophicephala (Pathansali, 1974)

Aidanosagitta regularis (Aida, 1897)

Aidanosagitta septata (Doncaster, 1903)

Aidanosagitta tropica (Tokioka, 1942)

Genus *Caecosagitta* Tokioka, 1965

Caecosagitta macrocephala (Fowler, 1904) (Fig. 11.12)

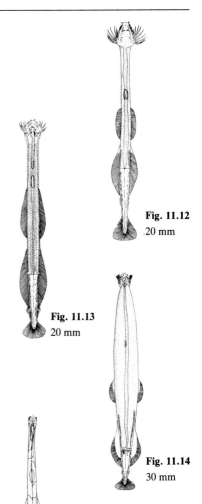

Genus *Ferosagitta* Kassatkina, 1971

Ferosagitta robusta (Doncaster, 1902) (non-Alvariño, 1962*b*) (Fig. 11.13)

Ferosagitta americana (Tokioka, 1959)

Ferosagitta ferox (Doncaster, 1902) (non-Alvariño, 1962*b*)

Ferosagitta galerita (Dallot, 1971)

Ferosagitta hispida (Conant, 1895)

Ferosagitta paulula Kassatkina, 1971 Syn. of *Zonosagitta pulchra*?

Ferosagitta tokiokai (Alvariño, 1967) Syn. of *F. robusta*? (Pathansali
  1974)

Genus *Flaccisagitta* Tokioka, 1965

Flaccisagitta hexaptera (d'Orbigny, 1836) (Fig. 11.14)

Flaccisagitta adenensis (Casanova, 1985)

Flaccisagitta enflata (Grassi, 1881)

Genus *Mesosagitta* Tokioka (1965)

Mesosagitta minima (Grassi, 1881) (Fig. 11.15)

Mesosagitta batava (Biersteker and van der Spoel, 1966)

Mesosagitta decipiens (Fowler, 1905)

Mesosagitta neodecipiens (Tokioka, 1959) Syn. of *M. decipiens*? (Pierrot-
  Bults 1979)

Mesosagitta sibogae (Fowler, 1906)

Genus *Parasagitta* Tokioka, 1965

Parasagitta elegans (Verrill, 1873) (Fig. 11.16)

Parasagitta euneritica (Alvariño, 1962)*

Parasagitta friderici (Ritter-Záhony, 1911)*

Parasagitta megalophthalma (Dallot and Ducret, 1969)

Parasagitta peruviana (Sund, 1961)*

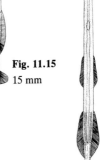

Fig. 11.12
.20 mm

Fig. 11.13
20 mm

Fig. 11.14
30 mm

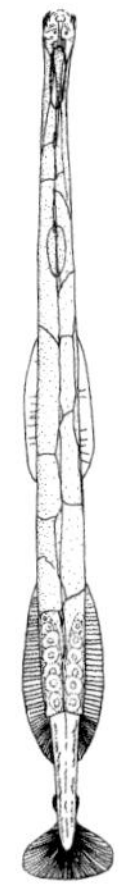

Fig. 11.15
15 mm

Fig. 11.16
30 mm

Table 11.2   *(continued)*

Parasagitta popovicii (Sund, 1961)*
Parasagitta setosa (Müller, 1847)
Parasagitta tenuis (Conant 1896)*

Genus *Pseudosagitta* Germain and Joubin, 1912
Pseudosagitta lyra (Krohn, 1853) (Fig 11.17)
Pseudosagitta gazellae (Ritter-Záhony, 1909)
Pseudosagitta maxima (Conant, 1896)
Pseudosagitta scrippsae (Alvariño, 1962) Syn of *P. lyra*? (Tokioka 1974*b*)

Genus *Serratosagitta* Tokioka, 1965
Serratosagitta serratodentata (Krohn, 1853)
Serratosagitta bierii (Alvarino, 1961)
Serratosagitta pacifica (Tokioka, 1940)
Serratosagitta pseudoserratodentata (Tokioka, 1939)
Serratosagitta selkirki (Faggetti, 1958) Syn. of *S. tasmanica*?
Serratosagitta tasmanica (Thomson, 1947) (Fig. 11.18)

Genus *Solidosagitta* Tokioka, 1965
Solidosagitta planctonis (Steinhaus, 1896) (Fig. 11.19)
Solidosagitta marri (David, 1956)
Solidosagitta zetesios (Fowler, 1905) Syn. of *S. planctonis*? (Pierrot-Bults
   1969)

Genus *Zonosagitta* Tokioka, 1965
Zonosagitta bedoti (Beraneck, 1895) (Fig. 11.20)
Zonosagitta bruuni (Alvariño, 1967) Syn. of *Z. littoralis*? (Tokioka 1959)
Zonosagitta izuensis (Kitou, 1966)
Zonosagitta littoralis (Tokioka and Pathansali, 1965)
Zonosagitta lucida (Casanova, 1985)
Zonosagitta nagae (Alvariño, 1967) Syn. of *Z. bedoti*? (Tokioka 1974)
Zonosagitta pulchra (Doncaster, 1902)

PTEROSAGITTIDAE Tokioka, 1965

Genus *Pterosagitta* (Costa 1869)
Pterosagitta draco (Krohn, 1853) (Fig. 11.21)

PTEROKROHNIIDAE nov. fam.

Genus *Pterokrohnia*
Pterokhronia arctica Srinavasan 1986 (Fig. 11.22)

* See Tokioka 1974 for syn.?

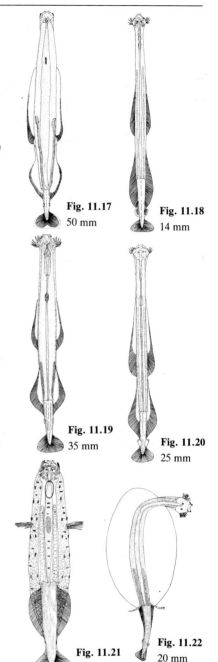

Fig. 11.17
50 mm

Fig. 11.18
14 mm

Fig. 11.19
35 mm

Fig. 11.20
25 mm

Fig. 11.21
12 mm

Fig. 11.22
20 mm

Table 11.3   Artificial key to the families of Chaetognatha

1a. Hooks but no teeth in the head. (If in doubt, clear in glycerine, mount under a cover glass and use 100×) 2
1b. Hooks and teeth present in head 3

2a. No eyes visible, ova mature asynchronously .... Krohnittellidae, 3 spp
2b. Eyes present, ova mature synchronously .... Bathyspadellidae, 1 spp

3a. One paired row of teeth in the head 4
3b. Two paired rows of teeth in the head 7

4a. Teeth are very conspicuous, 'flabelliform', with wide bases overlapped like a fan .... Krohnittidae, 2 (3?) spp
4b. Teeth are slightly wedge-shaped or long and narrow, not exceptionally broad at the base 5

5a. Teeth are curved and may wrap the hooks. No ventral ganglion, but a small dorsal ganglion just posterior to the head .... Bathybelosidae, 1 sp
5b. Teeth are straight or crooked, not curved. Ventral ganglion is prominent 6

6a. Teeth numerous, more than ten. Body long and narrow, moderately flexible to flabby. .... genus *Eukrohnia*, 10 spp, in the Eukrohniidae
6b. Teeth relatively few, less than ten. Body short and very stiff. .... Spadellidae, 18–19 spp.

7a. Two paired rows of teeth, two pairs of lateral fins (fins may be connected by a band of tissue or appressed in the genus *Pseudosagitta*) .... Sagittidae, 55–60 spp. See key to genera
7b. Two paired rows of teeth, one pair of lateral fins 8

8a. Lateral fins entirely on the tail segment or barely reach onto the trunk 9
8b. Lateral fins extend well onto the trunk approaching or reaching the ventral ganglion. .... Eukrohniidae, genera *Heterokrohnia*, 10–11 spp, and   *Archeterokrohnia*, 2–3 spp

9a. Lateral fins completely rayed, fins on tail segment only. Large cells of fragile collarette give appearance of wire mesh. Corona visible only with stain. .... Pterosagittidae, 1 sp
9b. Lateral fins usually completely rayed but may have small rayless zone next to body. Fins may extend slightly onto the trunk, alveolar tissue firm, of small cells. Corona visible without staining. .... Spadellidae, 18 spp

and defining maximal homogeneity in each category.

Clearly the genera of Tokioka contain much more information than the genus *Sagitta* of Ritter-Záhony (Table 11.5); the latter merely indicates a chaetognath with two pairs of lateral fins and two pairs of anterior and posterior teeth. In contrast, Tokioka's genus with the appropriate name, *Serratosagitta* carries the information of two sets of lateral fins, two sets of anterior and posterior teeth, hooks with large serra-

tions, seminal vesicles of unique and complex structure, and eyes of similar type—characters which uniquely set off this distinctive group within the family, Sagittidae. It is a genus composed of at least five closely related species (see Table 11.5) distinct from all others in the family.

If another valid function is to group relatively homogeneous objects, then Tokioka's genera fulfil this criterion much better than the genus *Sagitta* sensu Ritter Záhony, a very heterogeneous collection of species.

Table 11.4   An artificial key to the genera of Sagittidae (For mature, well preserved specimens only.)

Sagittidae—two paired rows of teeth, two pairs of lateral fins (fins may be appressed or joined by inflated tissue)

To observe fins accurately, stain in water soluble aniline blue or other stain, or use transmitted light and vary the angle of incidence. Fins may be partially or completely absent in damaged or poorly preserved specimens.

1a   Body with thin musculature. A needle beneath the specimen is easily seen. Body shape is tumid, inflated like a long balloon or sausage, with maximum body width near the middle of the specimen. Trunk 1½ to 2 times as wide as head 2

1b   Body stiff with strong musculature (specimen can be picked up in the middle without bending in half). If with thin musculature, the trunk is about as wide as the head, bean sprout-shaped, not balloon-shaped; maximum body width near the middle of specimen. 3

(a)   (b)

2a   Anterior fins have no connection to the posterior fins and are spaced far back from the ventral ganglion. The vestibular pit is circular (Use 40× or higher) .... *Flaccisagitta*, 3 spp

2b   Anterior fins come very close to or touch the posterior fins or are joined to the posterior fins by inflated tissue. Anterior fins come close to or reach the ventral ganglion. The vestibular pit, if visible, is a slit. .... *Pseudosagitta*, 3–4 spp.

3a   Inner edges of hooks have heavy, coarse serrations visible at 100× if the hook is flat to the plane of focus .... *Serratosagitta*, 5–6 spp

3b   Inner edges of hooks are smooth (may have fine denticulations visible at 400× or with SEM) 4

4a   Head is almost rectangular in shape, longer than wide, numerous fine, brown hooks. Body may have traces of yellow or orange pigment. Specimens are usually taken in tows deeper than 400 m. .... *Caecosagitta*, 1 sp

4b   Head not as above 5

5a   Both pairs of fins have a long, rayless zone next to the body or there is a rayless zone at the anterior end of both pairs of fins, or anterior fin has a few scattered or no rays 6

5b No rayless zone or if present it is narrow and does not reach all the way to the anterior end of both sets of fins 8

6a Gut diverticula absent (clear specimen in glycerine, lactose, or dissect at neck) ..... *Zonosagitta*, 5–7 spp

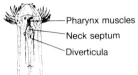

Pharynx muscles
Neck septum
Diverticula

6b Gut diverticula present 7

7a Body stiff with very strong muscles and conspicuous alveolar tissue. Usually mature at greater than 25 mm length .... *Solidosagitta*, 4 spp

7b Body musculature weak, alveolar tissue inconspicuous or missing. Usually mature at less than 22 mm .... *Mesosagitta*, 4–5 spp.

8a Anterior fin rays are sparse, perpendicular to the side of body or project forward. Small when mature, usually less than 10 mm seldom exceeding 15 mm .... *Aidanosagitta* 16–18 spp

8b Fin rays completely appressed, no sparse area. In both sets of fins, rays at the anterior ends of fins project backward. Almost always longer than 10 mm when mature, often exceeding 15 mm 9

9a Head wider than the body width, strong body muscles. Hooks usually thrown out to the side by the contraction of the powerful head muscles. Mature ovaries usually fill half or more of the length of the trunk 10

10a Gut diverticula absent .... *Sagitta*, 3 spp
10b Gut diverticula present .... *Ferosagitta*, 5–7 spp

The chaetognath species

By species is meant an interbreeding group of organisms that is reproductively isolated from other interbreeding groups. Hengeveld (1988) has taken exception to Mayr's recent (1988) redefinition of the biological species because it includes the concept of the species-specific niche. In view of the pioneering work of Paine (1966), showing that multiple species can occupy the same niche under conditions of heavy grazing or predation, Hengeveld is correct. This has particular significance in the evolution and ecology of the chaetognaths, where niche overlap seems so prevalent and heavy predation on chaetognaths is a possibility. The biological species concept should not include the concept of species-specific niche.

In applying the biological species concept to the Chaetognatha we have not only the problem of determining the status of disjunct populations, such as *Aidanosagitta regularis* in the eastern and western

Pacific, but also the problems of potential cloning due to self-fertilization, as demonstrated by Jägersten (1940) in *Parasagitta setosa*.

However, the greatest puzzle facing chaetognath systematists, a puzzle of over-riding importance to comparative morphologists, physiologists, developmental biologists and ecologists, is the problem of the extreme and abrupt morphological variation demonstrated in *Aidanosagitta crassa* (Tokioka 1938, 1939, 1965a) due to environmental conditions and elucidated experimentally in the unique work of Murakami (1957, 1959). If Tokioka, Murakami, Kado (1954) and Kado and Hirota (1957) are right, this abrupt and extreme morphological variability has important implications for many species pairs, the most famous being the *Parasagitta tenuis–P. friderici* pair, but also included are *P. peruviana–P. popovicii*, *Ferosagitta robusta–F. galerita* (Dallot 1971; Ducret 1973), *Zonosagitta*

Table 11.5 Classification of the genera in the family Sagittidae Claus & Grobben 1905 by different authors

	Ritter-Záhony (1911)	Tokioka (1965)	Bieri, this volume
Sagitta	*bipunctata*	*bipunctata*	*bipunctata*
	helenae	*helenae*	*helenae*
	setosa	*bombayensis*	*bombayensis*
	friderici	*setosa*	
	bedoti	*setosa var. euxina*	
	pulchra	*friderici*	
	serratodentata	*tenuis*	
	elegans		
	robusta (ferox)		
	neglecta		
	regularis		
	minima		
	decipiens (sibogae)		
	planctonis (zetesios)		
	macrocephala		
	hexaptera		
	enflata		
	lyra		
	gazellae		
	maxima		
Zonosagitta		*bedoti*	*bedoti*
		pulchra	*pulchra*
			bruuni
			izuensis
			littoralis
			lucida
			nagae
Serratosagitta		*serratodentata*	*serratodentata*
		pseudoserratodentata	*pseudoserratodentata*
		bierii	*bierii*
		pacifica	*pacifica*
		tasmanica	*tasmanica*
Parasagitta		*elegans*	*elegans*
		robusta	*setosa*
		ferox	*tenuis*
		hispida	*friderici*
			euneritica
			megalophthalma
			peruviana
			popovicii

Table 11.5 (*continued*)

	Ritter-Záhony (1911)	Tokioka (1965)	Bieri, this volume
Ferosagitta			robusta
			ferox
			hispida
			americana
			galerita
			paulula
			tokiokai
Aidanosagitta		neglecta	neglecta
		regularis	regularis
		bedfordii	bedfordi
		oceanica	oceania
		johorensis	johorensis
		demipenna	demipenna
		crassa	crassa
		delicata	delicata
		tropica	tropica
			alvarinoae
			erythraea
			guileri
			ophicephala
			septata
Mesosagitta		minima	minima
		decipiens	decipiens
		neodecipiens	neodecipiens
			sibogae
Solidosagitta		planctonis	planctonis
		zetesios	zetesios
		marri	marri
Caecosagitta		macrocephala	macrocephala
Flaccisagitta		hexaptera	hexaptera
		enflata	enflata
		gardineri	adenensis
		lyra	
		gazellae	
		maxima	
Pseudosagitta			lyra
			scrippsae
			gazellae
			maxima

bedoti–Z. littoralis, *Z. bruuni–Z. nagae* and others. Tokioka earlier (1959, 1962, 1965a) discussed the *P. friderici–tenuis* problem without reaching a definite conclusion, but in 1965b, he decided firmly that each species pair is a variant of a single interbreeding population. More recently, both McLelland (1980, 1984, 1989b) and Michel (1984) have argued for the existence of two distinct species. The restriction of *Ferosagitta galerita* to warm, high salinity inlets is very similar to the *Parasagitta tenuis* situation. The possibility that *Ferosagitta galerita* is a variant of *F. robusta* cannot be ruled out. However, until very fine-scale temporal and geographic field analysis or controlled laboratory experiments with living animals demonstrate conclusively that a single species is involved, the pragmatic solution to this problem discussed by Tokioka (1974a, 1974b, 1979) is to indicate the separate nature of the morphology by using different specific names.

In view of this highly variable morphology, discrimination between species can be difficult if not problematical. Tokioka (1974b) has summarized his views on which morphological characters are relatively stable (and thus usually good guides to species discrimination) and which must be viewed with caution. As a minimum, these seven characters should be discussed in any description of a new species:

1. General appearance of the body, including the proportional length of the tail segment, which can change with maturation rate; the stiffness of the musculature and the transparency of the body. General appearance also includes location of maximum body width, relative width of the head compared to body width, degree of taper of the tail segment, and degree of taper of the anterior portion of the trunk. Compare, for example, *Mesosagitta minima* and *Ferosagitta robusta* in Table 11.2.

2. The structure and arrangement of the lateral fins, including orientation of the rays and distribution of the rayless zone.

3. The position, shape, and length of the corona ciliata. This can change with preservation (Bieri 1974b).

4. Outline of the pigment cell of the eye. This may change with age in some species or vary in the same individual. Kotori and Kobayashi (1979) show the pigment shape in *Parasagitta elegans* as a 'T' in one eye and as a cross in the other eye. See also Ducret (1968) for variation in *Sagitta bipunctata*.

5. Structure and location of the seminal vesicle in respect of the tail fin and the posterior fin.

6. Armature formula which changes with body size.

7. Presence or absence of the intestinal diverticula. This character is not always definitive in view of its variation in different species (Dallot, 1970), and is often difficult to see in formalin-preserved species with moderate to heavy musculature. Kassatkina (1982) recommends dissection whereas Tokioka (personal communication) recommends clearing in lactophenol or glycerine.

I would add (although variable, as with armature formulas), extent of the mature ovary, disposition of the eggs within it, and whether the eggs mature synchronously or over a period of time.

Many of these characters can be quantified using various proportional measurements, the ratio of tail segment length to total length 'T' being most commonly cited. Unfortunately, no standardized system of comparison has been agreed upon. Most workers omit the tail fin from all length measurements because it is often damaged.

As morphological characters which can change markedly in response to different environmental conditions, Tokioka (1974b listed:

(1) size of mature individuals;

(2) relative length of the anterior fin;

(3) ratio of length of posterior fin on the trunk to length of posterior fin on the tail segment, although the ratio is related to the rate of maturation;

(4) the formation of fin rays which is accelerated by maturation;

(5) shape of the tail fin;

(6) more fully developed alveolar tissue at colder temperatures or at lower salinities;

(7) increased number of hooks with age, until near maturity when they may decrease in number (this may be complicated by temperature);

(8) increased number of anterior and posterior teeth with age, but this is affected by temperature.

Many workers have used tabular, graphical, and statistical techniques to study morphometric variation, for example, David (1956), Ducret (1968), Pereiro (1972), Dallot and Laval (1974), and Ibañez *et al.* (1974). The major limitation of such techniques is insufficient geographical coverage of the species and insufficient fine-scale sampling of important gradients such as the near shore, in lagoons and between regions of rapid hydrographic change. Despite these limitations, such analyses should be encouraged.

In reviewing the species listed in Table 11.2, a few species are well-defined, with little or no controversy related to their identity or taxonomy. Such species are *Pterosagitta draco*, *Krohnitta subtilis*, *K. pacifica*, *Caecosagitta macrocephala*, *Flaccisagitta hexaptera*, *Fl.*

enflata, and *Mesosagitta minima*. Interestingly, except for *Flaccisagitta* and *Mesosagitta*, all of these are either in mono- or dispecific genera. Nearly every other species listed has some question about taxonomic status, variability, geographic or depth range or has been so little seen by other workers that they remain almost unknown. Thirty-five of the species listed in Table 11.2 have been seen by only one scientist or have been collected only once! Even such often studied species as *Parasagitta elegans*, *P. setosa*, *Spadella cephaloptera*, *Paraspadella schizoptera*, and *Eukrohnia hamata* are not yet well described in terms of morphological variation over their known range, and for some the geographic range is poorly known. Space does not allow a more complete discussion of the many problems hidden in Table 11.2.

The chaetognath genera

If the species of Chaetognatha are in some disarray, the status of the genera is even more contentious. About 28 genera have been erected—rather a sizable number considering that at present about one hundred species are known. The first ten genera listed in Table 11.2 are relatively well established, except *Bathyspadella*, which is still known from only one specimen, thoroughly described by Tokioka; *Krohnittella*, which is known from only three specimens, two collected by Germain and Joubin (1916) and one by Bieri (1974*a*); and *Bathybelos*, which is known from a single specimen collected and thoroughly described by Owre (1973).

Although there are problems in the above genera, the division of the genus *Sagitta* (sensu Ritter-Záhony, 1911) into eleven genera is clearly the most contentious of the problems at the generic level. The recognition of 'groups' within the Sagittidae has been traced back at least as far as Abric (1905), but before Tokioka's revision, no one had tried to include all the known sagittids within well defined groups. Germain and Joubin (1916) discussed the groups *hexaptera*, *bipunctata*, and *macrocephala*—all extremely heterogeneous. Furnestin (1957) discussed the groups *friderici*, *bipunctata*, *serratodentata*, *lyra*, *hexaptera*, and *enflata*. She included ecologic and zoogeographic considerations in devising her groups, which are not morphologically homogeneous. She did not consider the Indo-Pacific species. Alvariño (1963) described eight groups, omitting such well known species as *Zonosagitta bedoti*, *Z. pulchra*, and *Caecosagitta macrocephala*. In 1967, using both morphological and ecological characteristics, she listed the groups *maxima*,

hexaptera, *serratodentata*, *euneritica*, *bipunctata*, *planctonis*, *bedoti*, *elegans*, *hispida*, *neglecta* and *oceania*. She left *Zonosagitta pulchra* out of these groups.

'Groups' have also been discussed in the genus *Spadella* s. Ritter-Záhony (Tokioka 1951; Bieri 1974*b*; Alvariño 1981) but were not given formal status until Salvini-Plawen (1987) briefly erected two new genera. Bowman and Bieri (1989) revised his genera, recognizing only *Spadella* and *Paraspadella* as valid. 'Groups' have also been discussed in the genus *Eukrohnia* (Casanova 1986*e*, McLelland 1988*a*) but no formal proposals have yet been made.

Tokioka's 1965*a* revision has not been widely accepted. His genera *Sagitta* and *Parasagitta* are heterogeneous and must have struck North Atlantic workers as unrealistic. Putting *elegans* into a group with *robusta* and *hispida*, while separating *setosa* into another group with *bipunctata* and *helenae* on the basis of a single character (the presence or absence of the gut diverticula) was neither convincing nor useful. Fifty-five years elapsed between the revisions of Ritter Záhony and Tokioka with only four new genera of questionable validity proposed in that time. The creation by Tokioka of eight new genera in a single paper essentially doubled the number of genera in common usage at that time and was somewhat overwhelming. Since Tokioka's revision, ten new genera have been proposed, five of which are included in Table 11.5. Five genera of Kassatkina (1982) are omitted. The extensive works of Kassatkina have not been included in this discussion except for the genus *Ferosagitta*. Kassatkina's papers can be traced back from her 1982 paper. The status of one family and two of her genera are discussed in Bieri (1991*a*). The remaining species and genera are not drawn or described clearly enough to permit taxonomic discrimination or identification.

The problems of these genera can be discussed conveniently in two stages. The first question is, 'Are there morphologically homogeneous groups of species within the family Sagittidae that are morphologically distinct from other groups?' The answer to this must be 'Yes'. Most conspicuous is the *serratodentata* group named *Serratosagitta* by Tokioka (1965*a*). Not only Germain and Joubin (1916) but also Furnestin (1957), Alvariño (1963) and Bieri (unpublished data) recognized the unique nature of this group. It has been repeatedly referred to by other scientists since Tokioka's revision (Pierrot-Bults 1974; Pineda-Polo 1979). The second group, almost universally recognized, is composed of the large, flabby, thin-walled and inflated species which have been placed in either one of two groups, the *hexaptera—enflata* group and the *lyra— maxima* group. Tokioka put these species into one genus, *Flaccisagitta*, but, agreeing with Furnestin and Alvariño, I prefer to keep the *lyra* group separate in the genus *Pseudosagitta* of Germain and Joubin. Of the other groups that have been proposed informally, the *neglecta* group (Tokioka's *Aidanosagitta*) seems relatively homogeneous and distinct. Tokioka's genera *Parasagitta* and *Sagitta* are clearly heterogeneous and there are smaller inconsistencies in *Zonosagitta* and *Aidanosagitta*. Further division of these genera seems necessary (Bieri 1991*a*) and as more detailed knowledge develops further changes may be required.

Since there are at least some groups within the Sagittidae that are morphologically homogeneous and morphologically distinct from other groups the second question is 'Should these groups be given formal status?' Tokioka has done this. I would argue that raising these groups to generic status is desirable for two reasons. First, if formal generic names are used, the species group must be given each time the trivial name is used and this immediately gives the reader much more information about the species morphology and about species which resemble it. This is enor-

mously useful, not only to the systematist but also to ecologists and to workers in other fields. For example, a worker in the Western Mediterranean has a number of specimens that seem to be, according to Ritter-Záhony's classification, *Sagitta friderici*. The question 'What other chaetognaths are similar to this species?' is answered by searching through Table 11.5. There *friderici* is listed in the genus *Parasagitta* and from the Table the investigator finds immediately that *friderici* is closely allied to seven other described species. (I ask the forbearance of the reader for the inclusion of the heterogeneous species in the genus *Parasagitta*. These species require further formal generic treatment inappropriate here). Similarly, an investigator finds the name *Sagitta guileri* in a recent paper. What does this chaetognath look like and what other species are allied to it? The species is in the formal group (genus) *Aidanosagitta*, with 13 other described species. This is much more helpful than trying to run the species down among the 55 to 65 species that would be listed under Ritter-Záhony's *Sagitta*. Furthermore Ritter-Záhony's listing gives no hint of morphology or of allied species.

Secondly, formal status should eventually decrease the number of changes between groups because formal changes require a more critical analysis and a formal published statement.

Although the taxonomy and systematics of the Chaetognatha is in a near chaotic state, there are some grounds for optimism. The diversity within the phylum is finally being recognized. If future revisions are more frequent and are carried out by families and by genera rather than by one investigator trying to work on the entire phylum, change will be more carefully done and the working scientist will be better able to comprehend and absorb the changes.

Note: Since the book has been in print, the following six new genera of Sagittidae have been proposed, with type specimens listed in parentheses: *Adhesisagitta (Sagitta hispida* Conant, 1895), *Demisagitta (Sagitta demipenna* Tokioka and Pathansali, 1963), *Decipisagitta decipiens (Sagitta decipiens* Fowler, 1905), *Tenuisagitta (Sagitta tenuis* Conant, 1896), *Abicasagitta (Sagitta pulchra* Doncaster, 1903) and *Oculosagitta megalophthalma* Dallot and Ducret, 1969) (Bieri 1991*c*).

Acknowledgements

I offer my sincere thanks to Q. Bone, T. E. Bowman, J-P. Casanova, J. A. McLelland, A. C. Pierrot-Bults, and E. V. Thuesen for help with the manuscript.

12

COLLECTION AND CHEMICAL ANALYSIS OF CHAETOGNATHS AND CHANGES DUE TO PRESERVATION

D V P Conway and D B Robins

Introduction

Chaetognaths are recognized as a key component in the marine food web (Reeve 1970a; Sameoto 1972; Kotori 1976; Szyper 1978; Hopkins 1982; Feigenbaum and Maris 1984; Angel 1989) but much of the basic information on the relationship between their length, weight and chemical composition is not generally available in sufficient detail for the study of energy movement through the web. Available data are scattered through the literature, are often based on very few measurements, and with few exceptions, are not detailed for any one species. In addition, the methods used are not standardized and are sometimes only superficially described. Most of this kind of information for marine zooplankton is available either for crustaceans (Childress and Nygaard 1974; Hirota 1981; Williams and Robins 1982; Böttger and Schnack 1986), fish larvae (Hay 1982, 1984) or simply for bulk zooplankton samples (Platt et al. 1969; Giguère et al. 1989). These studies provide useful supporting information and guidelines on techniques, but chaetognaths present special problems, because structurally they are an individual group (they are basically fluid-filled tubes with delicate walls) and require individual consideration.

In broad ecological studies, information on the contribution by chaetognaths to dry weight and chemical composition of the total zooplankton may not need to be precise, as it is pointless carrying out these analyses to a greater accuracy than the sampling strategies and methods themselves allow. Because of the difficulty of working immediately on large numbers of fresh specimens, chaetognath material is usually available chemically preserved in bulk plankton samples. The chaetognaths can be dry weighed in bulk and then, by using simple and accurate factors, estimates of fresh dry weight and chemical composition can be obtained. Different factors, if available, should be used for each species since a general 'chaetognath' factor may not be appropriate, as there can be considerable differences in chemical constituents between species (Bone et al. 1987a; Gorsky et al. 1988; Ikeda and Kirkwood 1989).

Greater accuracy is required for detailed experimental studies. This will involve either time-consuming individual dry weight measurement and chemical analysis of fresh chaetognaths, or individual length measurements of preserved material and application of accurate conversion factors or equations to these. The conversion must take into account the alteration in length which occurs during fixation and preservation, and the changing dry weight and chemical composition at different lengths. Chemical analysis is sometimes carried out on preserved material and these values are then converted to fresh values, but this naturally depends on a thorough knowledge of the effects of preservation. Even for fresh specimens, great care must be taken to ensure that the sampling methodology employed and the methods used to obtain the initial weight and chemical composition data are valid; it may be necessary to keep animals in aquaria, at least for short periods.

Collection and handling of chaetognaths

Pelagic chaetognaths are not difficult to collect, but there are problems in collecting samples that are truly representative of the population. *Sagitta elegans* for example, can show complex vertical distributions as a result of diurnal, seasonal and ontogenetic migrations (Conway and Williams 1986), and can be found concentrated close to the sea bed in shallow water (Jakobsen 1971; Oug 1977). Unless multiple depth or depth-integrated samples from very close to the bottom are taken, or separate epibenthic sampling is carried out, sampling may be biased.

Chaetognaths are often studied from samples taken during generalized zooplankton surveys and the equipment, towing speed and mesh size of net used may not be ideal. Damage to material in the cod-end can be lessened by using a finer mesh than the main net (Øresland 1987). It is unlikely that all stages can be effectively sampled with one mesh size. A species such as *Sagitta elegans* can be well sampled with a 200–300 μm mesh net, but the smaller specimens (and the eggs, which are approximately 330 μm (Kotori 1976)) may be extruded through the meshes. The youngest stages in the chaetognaths life cycle are very fragile and are usually destroyed on impact with coarse nets. The eggs and early stages, if in sufficient concentration, are better sampled with large volume water bottles and gently filtered onto fine netting. Tiselius and Peterson (1986) sampled *Sagitta elegans* with a 220 μm net and collected eggs by pumping water from selected depths and filtering known volumes onto 64 μm netting.

The type and complexity of net used to sample specifically for chaetognaths depends on the type of study (and ultimately on the resources and equipment available). Live material in best condition can be collected in containers by divers (Reeve 1977) or if the animals are abundant, by dipping a bucket at the surface. Unfortunately, these options are rarely available. However, chaetognaths can be collected in excellent condition with the simplest conical hand net with a non-filtering or reduced filtering cod-end. A cod-end made of plastic pipe fitted with filtering windows at the upper end (Furnestin 1976; Matthews and Hestad 1977) is a good option—material is drawn into the cod-end and does not accumulate above the cod-end collar of the main net, as happens with commonly used non-filtering cod-ends. A variety of large volume containers and plastic bags have been used as cod-ends (Sameoto

1972; Sweatt and Forward 1985b; Øresland 1987). Øresland (1987) further suggested that a closing cod-end would be preferable, to prevent water exchange during recovery. This could be an important consideration if sampling at specific depths in strongly temperature—or salinity-stratified waters. Reeve (1977, 1981) has developed relatively sophisticated nets with large volume cod-ends specifically for sampling chaetognaths and transporting them to the laboratory. The cod-ends can be used as aquaria, so there is no necessity to transfer material.

Collecting and handling techniques for live chaetognaths are the same as for all live zooplankton. They should be sampled on short, gentle tows so that material is in good condition and not too much is collected to affect condition and make sorting tedious and protracted. Sampling periods as short as 10 seconds have been used (Nagasawa 1984). A gentle sampling method is simply to suspend the net in the tidal or current flow (Sweatt and Forward 1985). The net should not be washed down until the plankton has been removed. There should be minimum handling of chaetognaths, they should be kept cool, separated from dead plankton and attempts made to minimize their exposure to light. Sorting and handling should be done in a cool or constant temperature room (Ikeda and Kirkwood 1989) to try and maintain them close to the temperature at which they were sampled, although there is evidence of an improved survival rate if the animals are held at a temperature a few degrees below sampling temperature (Feigenbaum and Maris 1984). It has been suggested that handling should always be done by dipping the chaetognaths with a beaker (Feigenbaum and Maris 1984), although wide bore pipettes are commonly used for sorting. Øresland (1987) sorted using a large spoon to minimize contact with the chaetognaths, as he considered them very susceptible to contact damage. Sorted chaetognaths which are to be used for experimental purposes are generally held in aquaria for at least a day (Reeve 1977) to allow them to stabilize and for any obviously weak individuals to be identified.

Although Murakami (1959, 1966) was the first worker to keep a pelagic chaetognath (*Sagitta crassa*) for an extended period (up to 106 days), Reeve (1970b) was the first to successfully culture one (*Sagitta hispida*) through complete cycles of development. These and other successful extended experimental studies on pelagic chaetognaths in aquaria (Dallot 1968; Reeve and Walter 1972; Kotori 1976; Nagasawa 1984) have used neritic species caught inshore at the surface. Apart from careful tending, success is probably due to the particular resilience of the species used, especially *S. hispida* and *S. crassa*. Some species certainly appear to be more tolerant of capture than others. In mixed samples *Sagitta setosa* appeared noticeably less dam-

aged than *S. elegans* (Bone *et al.* 1987*a*). Once captured, chaetognaths can be maintained in the laboratory or aboard ship, at least for short periods, with quite simple handling facilities and techniques. Reeve (1977) has summarized the techniques necessary for laboratory maintenance of zooplankton.

Dry weight, carbon, and nitrogen determinations on fresh chaetognaths

Once chaetognaths are available for processing they must be worked on immediately. As with weight and chemical determinations on any organisms, care must be taken to exclude individuals in poor condition, or with features which could bias results. Only chaetognaths with empty guts should be used (Gorsky *et al.* 1988; Ikeda and Skjoldal 1989), because analysis of individuals with even a single copepod in their gut could radically alter weight and chemical determinations. Chaetognaths containing obvious parasites should likewise be excluded.

Accurate weighing and subsequent chemical analysis of fresh chaetognaths presents many problems. Weighing techniques have usually involved an initial attempt at removal of surface sea water by rinsing in distilled water (Kotori 1976; Matthews and Hestad 1977; Nagasawa 1984). Using 50 g samples of mixed zooplankton Platt *et al.* (1969) concluded that removal of inorganic salts from bulk samples by immersion in 1 litre volumes of distilled water, for periods from 30 seconds to 1 hour, did not significantly affect carbon content. However, they did not measure the initial carbon content in untreated material. Omori (1978) carried out dry weight and chemical determinations on small numbers of chaetognaths and copepods. The organisms were placed on a vacuum apparatus (< 250 mm Hg vacuum) and rinsed with increasing volumes of distilled water for periods of always less than 15 seconds and usually less than 3 seconds. There was an increasing reduction in carbon, nitrogen, and dry weight content with increasing volume of rinse. Results from the bulk samples and individual groups of animals are not entirely comparable because the bulk samples in distilled water would be immersed in body fluid exudations and residual sea water, which may reduce the osmotic differential.

To investigate pretreatment techniques to remove adherent salt prior to dry weight and chemical analysis, samples of *Sagitta elegans* were collected in the Celtic

Sea in September 1981 using a 0.5 m, 280 μm mesh conical hand net. To minimize sample damage, the net (fitted with a non-filtering, plastic, 2 litre cod-end) was drifted in the tidal stream in the upper 10 m for < 5 minute periods. Only freshly caught chaetognaths in perfect condition were used. Large numbers were measured live, under cool conditions, in petri dishes of sea water and then removed using light forceps and treated in three different ways prior to dry weighing. The first set received no treatment, being removed from sea water and dried immediately. The second set was quickly dipped in distilled water and the third was lightly blotted over their surface. If a large drop of water remained on the end of the chaetognaths in either of the first two sets it was lightly removed by blotting. A quick dip into distilled water should be an efficient way of removing salt. Distilled water remaining on the exterior may draw out internal body fluids because of osmotic differences, but these fluids will dry on to the surface and will not be lost to weighing.

Organisms for chemical analysis generally have to be dried and weighed first and it may be advantageous, at least for small specimens, if they can be made available on their own (and not, for example, in an aluminium boat or on a filter paper) as the dry weight of the organism may be a fraction of the weight of the carrier. Any slight weighing error could be more than the weight of the organism. When dried, Chaetognaths either stick to or leave deposits on the base of their carrier, whatever the material. To overcome this problem the chaetognaths were measured and suspended on thin wire wound over separately identified, compartmented plastic trays (Conway and Williams 1986); they were then dried in an oven. With care and practice, and by moving the chaetognaths slightly at periods during the early drying process, this technique provides individual chaetognaths with minimal sample loss and was used successfully for *Sagitta elegans* up to 22 mm in length. With chaetognaths longer than this the amount of body fluid would be such that it could drip onto the tray bottom during drying and be lost to analysis.

For very large chaetognaths, where there is not the problem of specimen weight relative to carrier weight, the best processing method is to place them in a small, precombusted, preweighed aluminium boat in which they can be wet weighed, dry weighed and then subsequently analysed chemically (Heron *et al.* 1988).

A variety of dry weighing temperatures have been investigated and used for zooplankton (Båmstedt 1974; Heron *et al.* 1988) but a temperature of 60°C (Lovegrove 1966) for 24 hours was used, as this is the most widely employed procedure. The duration of drying may have to be varied, depending on the size of chaetognath and the number, if being dried in bulk. As an indication of how rapidly organisms are dried, after drying for one hour at 70°C, mysids lose over 99.5 per cent of their water content (Båmstedt 1974). After drying, the chaetognaths were stored in a dessiccator until weighed on a Cahn 25 electrobalance.

The regression equations for length/dry weight, carbon, and nitrogen for the three treatments are given in Table 12.1. As expected, chaetognaths which received no treatment were slightly heavier than the other two treatments over their whole length range and also tended to have the highest carbon and nitrogen content. Chaetognaths which were blotted tended to have the lowest dry weight, carbon, and nitrogen content of the three treatments. Previously published regressions of length/dry weight for several species of chaetognath are summarized in Table 12.2 for comparison. When the carbon and nitrogen as a percentage of the dry weight is considered for the three treatments, separated into three size classes (Table 12.3), some differences are obvious. For the three treatments percentage carbon increased with increasing length class. In the untreated chaetognaths the percentage nitrogen increased with increasing length, while in the other two treatments it stayed roughly the same. The percentage carbon and nitrogen in the smallest size class of untreated chaetognaths is very low, which may be related to greater weight of salt retained as a result of their greater surface area to volume ratio. An increasing percentage carbon of dry weight with increasing length class, while percentage nitrogen stays in similar proportions, has also been shown for *Sagitta gazellae* (Ikeda and Kirkwood 1989), although the proportions of various elements in this species differ from most others. In general, the percentage of the dry weight which was carbon decreased from untreated through dipped to blotted, indicating a progressive loss of material, while percentage nitrogen showed no obvious trend.

When the C:N ratios of the three treatments are considered (Fig. 12.1), the dipped and blotted increase with increasing length and the untreated decrease with increasing length. It thus appears that while carbon and other materials are lost due to dipping and blotting, there is preferential loss of nitrogen.

It is known that *Sagitta elegans* has large vacuolated gut cells in the coelomic cavity, containing fluid rich in NH_4^+ and free amino acids (Bone *et al.* 1987a). This fluid could be susceptible to being drawn out by the processing methods. It is not known whether the small proportion of chaetognath species which have these vacuoles (Dallot 1970) also retain nitrogen-rich contents, but if they do it could cause interspecific differences in chemical analysis after rinsing in a hypotonic solution.

Table 12.1 *Sagitta elegans*. Length (mm)/dry weight (μg), carbon (μg), and nitrogen (μg) regression equations for chaetognaths untreated, dipped in distilled water, or blotted, before drying and chemical analysis. Number of observations (*n*) also given

	Untreated	Dipped	Blotted
Length/Dry weight	$W = 0.1595L^{3.08}$	$W = 0.1951L^{2.99}$	$W = 0.1114L^{3.2}$
n	167	239	232
Length/Carbon	$C = 0.0290L^{3.32}$	$C = 0.0473L^{3.14}$	$C = 0.03271^{3.24}$
n	50	76	65
Length/Nitrogen	$N = 0.0056L^{3.47}$	$N = 0.0197L^{3.00}$	$N = 0.0131L^{3.11}$
n	50	76	65

Table 12.2 Length (mm)/dry weight (µg) regressions from the literature for various chaetognath species converted to the same equation. Number of observations (*n*) on which the regression equations are based also given

Species	Length/dry weight regression	*n*	Reference
Sagitta elegans	$W = 0.064\ L^{3.30}$	19	Matthews and Hestad (1977)
	$W = 0.97\ L^{2.36}$	53	Sameoto (1971)
	$W = 0.206\ L^{2.85}$	33	Kotori (1976)
S. crassa	$W = 0.197\ L^{3.01}$	35	Nagasawa (1984)
	$W = 0.741\ L^{2.34}$	100	Hirota (1981)
Eukrohnia hamata	$W = 0.320\ L^{3.00}$	109	Matthews and Hestad (1977)
E. bathypelagica (Stage 0–II)	$W = 0.072\ L^{3.20}$	10	Matthews and Hestad (1977)
E. bathypelagica (Stage III–IV)	$W = 5.700\ L^{2.20}$	10	Matthews and Hestad (1977)

Table 12.3 *Sagitta elegans*. Comparison of the percentage carbon and nitrogen of the dry weight for chaetognaths which were untreated, dipped in distilled water, or blotted, separated into three size classes

Length range (mm)	Mean %N	Mean %C	*n*
Untreated			
<10	7.67 ± 2.3	28.22 ± 5.7	16
10–15	10.72 ± 1.3	37.52 ± 2.7	16
>15	11.71 ± 0.8	39.45 ± 1.3	18
Total	10.12 ± 2.3	35.24 ± 6.1	50
Dipped			
<10	10.19 ± 1.53	32.76 ± 3.1	23
10–15	11.13 ± 1.2	36.45 ± 2.7	25
>15	10.45 ± 0.8	37.76 ± 2.2	28
Total	10.59 ± 1.2	35.81 ± 3.4	76
Blotted			
<10	10.04 ± 1.8	31.99 ± 3.4	23
10–15	9.45 ± 1.5	33.45 ± 3.5	24
>15	10.50 ± 0.8	36.17 ± 3.4	18
Total	9.95 ± 1.5	33.69 ± 3.8	65

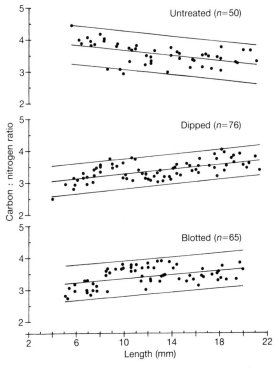

Fig. 12.1 *Sagitta elegans*. Length/carbon: nitrogen ratio regression plots for untreated chaetognaths, and chaetognaths dipped into distilled water or blotted before analysis. The 95 per cent prediction limits of the data and number of observations (*n*) are also shown.

There is thus a quandary in knowing whether to treat chaetognaths prior to dry weight and chemical analysis. While the dry weight in untreated specimens is probably slightly too high because of surface salt, the actual amount of salt in sea water is approximately 3.5 per cent by weight, so any weight increase is going to be small. If the chaetognaths are processed by dipping or blotting, material is drawn out, altering the dry weight and chemical composition, so it is probably best to always work on untreated chaetognaths. At least these are chemically intact! The commonly used techniques of measurement of dry weight and chemical composition may thus only provide accurate estimates, there being no way to measure them absolutely.

The speed with which materials are extracted from the chaetognaths during a rapid distilled water dip confirms the observations of Omori (1978). Obvious signs of osmotic stress can be seen after a few seconds when freshly caught *Sagitta elegans* are placed in distilled water and observed under a microscope. The whole animal shrinks, turns opaque and the body wall may rupture in several places. Sperm, if present, may be evacuated, vacuolated gut cells burst and fluid emerges from the buccal and anal regions. Not all specimens react at the same rate or in the same way.

Any length/dry weight/chemical composition conversion factors or equations obtained are unfortunately only applicable with reasonable accuracy to the particular material examined. As Omori (1969), Kankaala and Johansson (1986), and Ikeda and Kirkwood (1989) point out, apart from differences between closely related species, there can be great variations in weight and chemical composition for the same length of organism as a result of differences in nutritional and reproductive status and age. There can be seasonal and geographical variations and variations under different environmental conditions. For example, from experimental results, the biochemical composition of *Sagitta hispida* can fluctuate widely over a year (Reeve *et al.* 1970), suggesting that this was due to environmental parameters other than changes in food availability. These variations in weight and chemical composition indicate the caution that must be exercized when using conversion factors.

Wet weight determinations on chaetognaths

Because chaetognaths have a greater water content than most of the other dominant zooplankton groups (Ikeda and Skjoldal 1989), when only the dry weight is considered as a measure of biomass their ecological impact can be underestimated. The water content of a selection of chaetognaths has been summarized by Ikeda and Kirkwood (1989) and ranges from 83–94.7 per cent of the wet weight. The value for *Sagitta elegans* was 91 per cent (Båmstedt 1981). Because of the delicate structure of chaetognaths, and the amount of fluid they contain, wet weight measurement is extremely difficult to standardize (Beers 1976). They are generally weighed in preweighed aluminium boats in which they can subsequently be dry weighed. If they are blotted (unrinsed or after a rapid rinse in distilled water) it should be done quickly and carefully because of the ease with which internal fluids are drawn out through the body walls, thus altering the final weight. Handling and weighing must be carried out rapidly to preclude loss due to evaporation. With care and careful standardization of the procedure, acceptably accurate results should be obtained, and these will at least be internally compatible. Another wet weighing method is to weigh the animals in sealed tared vials of sea water (Båmstedt 1981), but this involves similar problems.

Effects of preservation on dry weight, carbon, and nitrogen content of chaetognaths

Formaldehyde preservation of zooplankton can result in rapid loss of organic material (Lasker 1966; Omori 1978; Böttger and Schnack 1986) which, unless it is predictable, limits use of the material for chemical analysis. The range of formaldehyde based fixation/preservation mixtures in common use for general zoo-plankton is quite limited. The three that have been compared here are 4 per cent borax buffered formaldehyde (2 g of borax to 98 ml of 40 per cent formaldehyde) in sea water (salinity 34.9%) (SF), fresh water (FF) (fresh water was used rather than distilled for consistency of availability), and 2 per cent borax buffered formaldehyde in sea water (as above) with propylene glyol and propylene phenoxitol (GF) (solution No 16, Steedman 1976a).

Because marine organisms live in osmotic balance with sea water it has been suggested that they would be best preserved in an iso-osmotic preservation mixture. As a guide, an osmotic pressure (in milliosmoles) given by Steedman (1976a) for sea water was 1023, for 4 per cent formaldehyde in distilled water the osmotic pressure was 1462, for 4 per cent formaldehyde in sea water it was 2500 and for 2 per cent formaldehyde/propylene glycol/propylene phenoxitol mixture in sea water it was 2735. While 4 per cent formaldehyde in sea water and 2 per cent formaldehyde/glycol mixture have double the osmotic pressure of sea water, material still appears to be preserved in excellent condition when they are used. This observation is in line with work on the ultrastructure of marine organisms, where best results are usually achieved when the osmolarity of the fixative vehicle is similar to that of the body fluid rather than the total osmolarity of the fixative, i.e. the contribution to fixative osmolarity by the formaldehyde can be neglected (Bone and Ryan 1972).

Large samples of *Sagitta elegans* collected in September 1981 were measured fresh and then individually preserved for a total of 106 days in the three formaldehyde mixtures. They were then measured (see p. 144), dipped in distilled water, gently blotted and dried at 60°C on the base of plastic compartmented trays. Preserved chaetognaths do not generally exude contents or adhere to containers as fresh ones do. Chaetognaths preserved in GF, especially the larger specimens, occasionally left oily traces of the propylene compounds on the base of the trays as this is not driven off and can ooze out during drying. Those in FF lost approximately 75 per cent of dry weight compared to fresh untreated material, while those in SF lost only 50 per cent (Fig. 12.2). The greater loss in dry weight when FF is used is at first sight surprising, as it has an osmotic pressure closer to sea water than the other mixtures. This difference possibly relates to differences in rate of fixation of the tissues. *S. elegans* preserved in GF is a dry weight closest to fresh, but this is artefactual because of the retention of the propylene compounds. Use of GF cannot be recommended if the samples are to be chemically analysed because of the amount of carbon it introduces. Length/dry weight regressions for the different mixtures are given in Table 12.4.

It is interesting to compare the proportions of carbon and nitrogen in fresh and differently preserved *Sagitta elegans* (Fig. 12.2, Table 12.4). While the dry weight of

chaetognaths preserved in SF is approximately 40 per cent greater than those preserved in FF, the proportion of the dry weight which is carbon in those preserved in FF is 30 per cent higher than those in SF. The difference in carbon levels is attributable to the loss in dry weight. In the chaetognaths preserved in FF the loss in dry weight appears to be greater in non-carbon based material. The C:N ratios of the preserved samples are higher than the unpreserved material so loss of nitrogen is higher than loss of carbon during fixation and preservation. Omori (1978) also observed differential loss of nitrogen when the chaetognath *Sagitta nagae* was preserved in 4 per cent sea water formaldehyde. The C:N ratio changed from 3.2 to 4.2.

Giguère *et al.* (1989) have reviewed the effects of formaldehyde preservation on the dry weight of zooplankton to demonstrate the wide-ranging changes. The changes measured here were done on individual specimens in controlled tests. Chaetognath material from a zooplankton survey would be from samples varying in species composition and bulk (from predominantly oily crustaceans to a sparse collection of coelenterates). The interstitial fluid of each sample would have a different osmotic pressure and, because of the delicate structure of chaetognaths, this would have an unpredictable effect on their dry weight and chemical composition. This could make any analysis on preserved material from different samples potentially very inaccurate and not to be recommended, except for very broad studies where accurate estimates are not necessary.

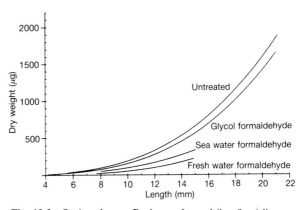

Fig. 12.2 *Sagitta elegans*. Back transformed ($\log_e/\log_e$) linear regressions of length/dry weight for chaetognaths, unpreserved or preserved in 4 per cent sea water formaldehyde, 4 per cent fresh water formaldehyde or 2 per cent glycol formaldehyde.

Table 12.4   *Sagitta elegans*. Carbon:nitrogen ratio and mean carbon and nitrogen content as a percentage of the dry weight in treated and fresh untreated chaetognaths. Length ranges and standard deviation of the means are also shown. Regression equations of length (mm)/dry weight (μg) carbon and nitrogen are also given for each mixture

	Untreated	4% FF*	4% SF†	2% GF‡
Length range (mm)	5.6–21.1	7.7–14.7	8.8–15.0	7.2–21.0
n	50	25	25	23
Carbon % of dry weight	35.2 ± 6.1	45.8 ± 3.7	33.5 ± 5.4	34.8 ± 3.5
Length/carbon regression (μm)	$C = 0.0290 \, L^{3.32}$	$C = 0.0683 \, L^{3.56}$	$C = 0.0225 \, L^{3.13}$	$C = 0.0158 \, L^{3.40}$
Nitrogen % of dry weight	10.1 ± 2.3	10.4 ± 1.1	7.3 ± 1.2	7.2 ± 1.1
Length/nitrogen regression	$N = 0.0056 \, L^{3.47}$	$N = 0.0016 \, L^{3.54}$	$N = 0.0025 \, L^{3.41}$	$N = 0.0017 \, L^{3.65}$
Carbon:nitrogen ratio	3.6 ± 0.3	4.4 ± 0.2	4.6 ± 0.3	4.9 ± 0.6
Length range (mm)	5.6–21.1	7.7–14.9	8.0–15.0	4.2–21.0
n	167	100	90	195
Length/dry weight regression	$W = 0.1595 \, L^{3.08}$	$W = 0.0119 L^{3.64}$	$W = 0.0847 L^{3.06}$	$W = 0.0985 L^{3.20}$

* 4% FF, borax buffered 4% formaldehyde in fresh water; † 4% SF, borax buffered 4% formaldehyde in sea water; ‡ 2%GF, 2% glycol formaldehyde in sea water; *n*, number of observations

The effects of fixation and preservation on chaetognath length

Rather than make dry weight and chemical measurements on preserved chaetognaths, it would be much simpler to measure their length and apply a factor or equation to make allowances for shrinkage due to fixation and preservation. These lengths could then be converted to unpreserved dry weight and chemical values. The feasibility of this has been investigated, again for the three common preservative mixtures.

Samples of freshly caught *Sagitta elegans* in perfect condition were measured and then individually placed in SF, FF and GF in marked compartmented trays. They were remeasured after 5 and 106 days, apart from those in GF which were only remeasured after 106

days. Of a sample of 95 *S. elegans* preserved in SF, after 5 days 96.8 per cent had shrunk by a mean of 10.3 ± 5.4 per cent. After 106 days all had shrunk by a total of 10.4 ± 5.2 per cent within the same shrinkage range. Thus, final length was virtually reached within 5 days. Percentage shrinkage increased slightly with length (Fig. 12.3). These percentage shrinkages are higher than found by Øresland (1985) who tested shrinkage in 30 *Sagitta elegans* (15–31 mm) in a similar sea water formaldehyde mixture, although the salinity of the sea water was probably less. Specimens were measured after 2 and 7 days and 30 months. Shrinkage varied between 3.7 and 10.3 per cent, with a mean of 6.3 per cent. Øresland found that no further shrinkage occurred after 2 days. Mean shrinkage of 5.4 per cent in a sample of 16 *Eukrohnia hamata*, in an unspecified formaldehyde mixture, was reported by Sands (1980).

A sample of 150 *Sagitta elegans* preserved in FF gave a very different picture (Fig. 12.3). After 5 days 5 per cent were unchanged in length while 49 per cent had shrunk by a mean of 10.7 ± 4.9 per cent. The remaining 46 per cent increased in length by a mean of 6.4 ± 3.2 per cent. Change in length was over a range of 39.2 per cent. After 106 days, 8 per cent were the same length as originally and 55 per cent had shrunk by a mean of 10.9 ± 5.8 per cent. The remaining 37 per cent had increased in length by a mean of 4.3 ± 2.7 per cent. Change in length was over a range of 31.1 per cent. There was thus a slight reduction in length overall between 5 and 106 days and the range of change diminished. There was also a slight reduction in shrinkage with increasing length. Change in length for any size was unpredictable and this was only demonstrable by examining individual length changes. If the chaetognaths had only been measured as a group, a mean shrinkage of 2.3 per cent would have been recorded after 5 days and of 4.5 per cent after 106 days.

A sample of 129 *Sagitta elegans* were preserved for 106 days in GF. All shrank with a mean shrinkage of 11.1 ± 4.6 per cent and there was only a slight increase in percentage shrinkage with increasing length (Fig. 12.3). Range of shrinkage was generally less than 20 per cent, although one individual shrank by 30.7 per cent. While mean shrinkage in SF and GF were quite close there was a greater increase in shrinkage with increasing length in SF. This may be related to speed of fixation, as propylene glycol may aid penetration of fixative (Steedman 1976*b*).

The differences between these three preservative mixtures have important implications for studies where measurements are being made on preserved chaetognaths. The data obtained give guidance on shrinkage, but they were individually made on carefully sampled, perfect specimens. During routine surveys, material would not be in as good a condition and would be preserved in bulk samples varying in osmotic pressure, which would make shrinkage unpredictable.

Formaldehyde prepared with fresh water cannot be recommended for studies which involve measurements, because of the very unpredictable and broad shrinkage response which only comes to light when individual length changes are recorded. Glycol formaldehyde preservation is probably the best of the three treatments tested if the chaetognaths are only to be meas-

ured and not chemically analysed. Any shrinkage factors obtained will not be interchangeable between species because of variation in rigidity of the body and length volume relationships etc.

To investigate the effects of osmotic pressure on shrinkage in more detail, samples of *Sagitta elegans* collected in the southern North Sea in June 1989 were measured fresh, then preserved in 4 per cent borax buffered formaldehyde mixtures prepared in a range of osmotic concentrations with sea water (33.4‰), diluted with distilled water in different proportions (Table 12.5). The chaetognaths were remeasured after 10 days.

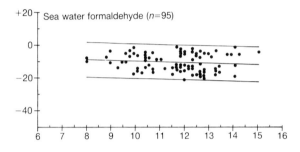

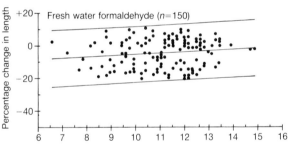

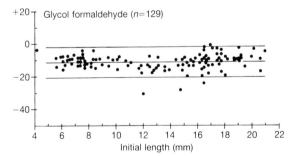

Fig. 12.3 *Sagitta elegans.* Percentage change in length from initial length after 106 days preservation in three different mixtures, 4 per cent sea water formaldehyde, 4 per cent fresh water formaldehyde or 2 per cent glycol formaldehyde. The regression lines, 95 per cent prediction limits of the data and number of observations (*n*) are also shown.

Table 12.5 *Sagitta elegans*. Percentage change in length after preservation in 4% borax buffered formaldehyde mixtures made up in a range of osmotic concentrations using sea water (33.4‰) diluted with distilled water in different proportions. The final mixture used distilled water alone. Length ranges of chaetognaths and number of observations (*n*) are also shown.

	Per cent sea water				
	100	75	50	25	Distilled water
n	24	24	24	23	22
	Per cent change				
unchanged	0	33	4	0	0
shrunk	100	42	13	9	14
lengthened	0	25	83	91	86
Length range (mm)	6.4–13.9	6.6–11.7	6.7–13.9	6.2–14.4	6.9–13.4

All shrank in the 100 per cent sea water formaldehyde, but in dilutions of sea water of 50 per cent and more, practically all increased in length. Mean shrinkage in 100 per cent sea water was 11.9 ± 3.1 per cent and even though the salinity of the sea water used was slightly less than for September 1981, the shrinkage was slightly higher. However, the number of chaetognaths measured from June 1989 was much smaller than the earlier 1981 sample of Fig. 12.3. There may be an osmotic pressure at which minimum length change takes place but this would be extremely difficult to regulate in bulk samples.

While chaetognaths are very different structurally and chemically from fish larvae, Hay (1981) has carried out some detailed experiments on the effects of handling and of different salinities and concentrations of formaldehyde on shrinkage in herring larvae which are worth considering here. As with *Sagitta elegans*, larval herring shrinkage increased with increasing osmotic pressure of preservative solution, although shrinkage was measured in groups, not individually. Laboratory reared larvae, measured and experimentally introduced into a plankton net, shrank during normal towing of the net and shrinkage was further greatly increased if preservation was delayed, giving higher shrinkage values than if they had been preserved directly. Greater shrinkage also occurred if larvae were killed before preservation. Whether these effects would also be applicable to chaetognaths has not been investigated, but they emphasize the problems which could be introduced during sampling and preservation.

If measured using accurate techniques, factors and equations to convert length and preserved dry weight to fresh dry weight and chemical values may enable acceptable estimates to be made. When errors due to unknowns in sampling efficiency, patchiness effects, etc. are considered, the inaccuracies introduced are probably quite acceptable, especially for broad energy flow studies. For detailed experimental studies it will still be necessary to carry out individual analyses on fresh animals.

REFERENCES

Abric, P. (1905). Sur la systématique des Chétognathes. *Compte rendu des Séances de la Société de Biologie, Paris*, **141**, 222–4.

Afzelius, B. (1969). Filament symmetries in muscles of the arrow-worm. In *Symmetry and function of biological systems at the macromolecular level* (eds A. Engström and B. Strandberg), Notel Symposium 11, pp. 415–21. Almqvist & Wiksell, Stockholm.

Ahnelt, P. K. (1980). Das Coelom der Chaetognathen. Unpublished Ph.D. Thesis. University of Vienna.

Ahnelt, P. K. (1984). Chaetognatha. In *Biology of the Integument, 1, Invertebrates* (ed. J. Hahn, A. G. Matoltsy and K. S. Richards), pp. 746–55. Springer Verlag, Berlin.

Aida, T. (1897). The Chaetognatha of Misaki Harbor. *Annotationes zoologicae japonenses*. **1**, 13–21.

Almeida Prado, M. S. de (1961). Chaetognatha encontrados em águas brasilieras. *Boletim do Instituto oceangrafico, São Paulo*, **11**, 31–55.

Almeida Prado, M. S. de (1968). Distribution and annual occurrence of Chaetognatha of Cananeia and Santos coasts (São Paulo, Brazil). *Boletim Instituto oceanografico São Paulo*, **17**, 33–55.

Alvariño, A. (1961). Two new chaetognaths from the Pacific. *Pacific Science*, **15**, 67–77.

Alvariño, A. (1962a) Two new Pacific chaetognaths, their distribution and relationship to allied species. *Bulletin of the Scripps Institution of Oceanography*, **8**. 1–50.

Alvariño, A. (1962b). Taxonomic revision of *Sagitta robusta* and *Sagitta ferox* Doncaster, and notes on their distribution in the Pacific. *Pacific Science*, **16**, 186–201.

Alvariño, A. (1963). Quetognatos epiplanctonicos del Mar de Cortes. *Rivista de la Sociedad Mexicana de Historia natural*. **24**, 97–203.

Alvariño, A. (1964). Bathymetric distribution of chaetognaths. *Pacific Science*, **18**, 64–82.

Alvariño, A. (1965). Chaetognaths. *Oceanography and Marine Biology, Annual Review*, **3**, 115–94.

Alvariño, A. (1967). The Chaetognatha of the NAGA Expedition (1959–1961) in the South China Sea and the Gulf of Thailand, Part 1: Systematics. *NAGA Report*, **4**, 1–197.

Alvariño, A. (1968). Egg pouches and other reproductive structures in pelagic Chaetognatha. *Pacific Science*, **22**, 488–92.

Alvariño, A. (1969). Los Quetognatos del Atlantico. Distribucion y notas esenciales de sistematica. *Trabajos del Instituto Español de Oceanografia*, **37**, 1–290.

Alvariño, A. (1970). A new species of *Spadella* (benthic chaetognatha). *Studies on the fauna of Curaçao and other Caribbean islands*, **34**, 73–89.

Alvariño, A. (1974). The importance of the Indian Ocean as origin of the species and biological link uniting the Pacific and Atlantic Oceans. *Journal of the Marine Biological Association of India*, **14**, 713–22.

Alvariño, A. (1978). *Spadella gaetoni*, a new benthic chaetognath from Hawaii. *Proceedings of the Biological Society of Washington*, **91**, 650–7.

Alvariño, A. (1981). *Spadella legazpichesi*, a new benthic chaetognath from Enewetâk, Marshall Islands. *Proceedings of the Biological Society of Washington*, **92**, 107–21.

Alvariño, A. (1983a). Chaetognatha. In *Reproductive biology of invertebrates. 1. Oogenesis* (ed. K. G. Adiyodi and R. G. Adiyodi), pp. 585–609. Wiley, London.

Alvariño, A. (1983b). Chaetognatha. In *Reproductive biology of invertebrates. 2. Spermatogenesis and sperm function* (ed. K. G. Adiyodi and R. G. Adiyodi), pp. 531–44. Wiley, London.

Alvariño, A. (1983c). The depth distribution, relative abundance and structure of the population of the chaetognath *Sagitta scrippsae* Alvariño 1962, in the California current off California and Baja California. *Anales Instituto Ciencias Mar y Limnologia Universitaria Nacional Autonomica Mexico*, **10**, 47–84.

Alvariño, A. (1987). *Spadella pimukatharos*, a new benthic chaetognath from Santa Catalina Island, California. *Proceedings of the Biological Society of Washington*. **100**, 125–33.

Andreu, P., Marasse, C., and Bardalet, E. (1989). Distribution of epiplanktonic Chaetognatha along a transect in the Indian Ocean. *Journal of Plankton Research*, **11**, 185–92.

Angel, M. V. (1979). Zoogeography of the Atlantic Ocean. In *Zoogeography and diversity of plankton* (ed. S. van der Spoel and A. C. Pierrot-Bults), pp 168–90. Bunge Utrecht.

Angel, M. V. (1986). Vertical migrations in the ocean realm: possible causes and probable effects. In *Migration: mechanisms and adaptive significance*. (ed. M. A. Rankin). Contributions Marine Science, Supplement. Volume 27. 45–70.

Angel, M. V. (1989). Vertical profiles of pelagic communities in the vicinity of the Azores front and their implications to deep-sea ecology. *Progress in Oceanography*, 22, 1–46.

Apstein, C. (1911). Résumé des observations sur le plancton des mers explorées pendant les annees 1902–1908. Pt. II Chaetognathes. *Bulletin trimestriel des Résultats aquis pendant les Croisières périodiques. Conseil permanent international pour l'Exploration de la Mer, Copenhague*, 2, 170–5.

Ass, M. Ya. (1961). The developmental cycle of nematodes of the genus *Contracae cum. Trudy Karadagskoi Biologicheskoi Stantsii*, 17, 110–2. (Translation in Fisheries Research Board of Canada. Translation series, No.530)

Bainbridge, V. (1963). Continuous plankton records: Contribution towards a plankton atlas of the North Atlantic and the North Sea. *Hull Bulletin of Marine Ecology*, 6, 40–51.

Baldasseroni, V. (1914). I Chetognati raccolti in Adriatico dalla 'R. N. Ciclope' nelle crociere III e VII. *Memori Rendi Compti Talassografica Italiano*, 38, 3–22.

Båmstedt, U. (1974). Biochemical studies on the deepwater pelagic community of Korsfjorden, western Norway. Methodology and sample design. *Sarsia*, 56, 71–86.

Båmstedt, U. (1981). Water and organic content of boreal macrozooplankton and their significance for the energy content. *Sarsia*, 66, 59–66.

Båmstedt, U. (1988). The macrozooplankton community of Kosterfjorden, western Sweden. Abundance, biomass and preliminary data on the life cycles of dominant species. *Sarsia*, 73, 107–24.

Beauchamp, P. de (1960). Classe des chétognathes (Chaetognatha). In *Traité de Zoologie 5* (ed. P. Grassé) (vol. 2), pp. 1500–20. Masson, Paris.

Beers, J. R. (1966). Studies on the chemical composition of the major zooplankton groups in the Sargasso Sea off Bermuda. *Limnology and Oceanography*, 11, 520–8.

Beers, J. R. (1976). Determination of zooplankton biomass. Gravimetric methods. In *Zooplankton fixation and preservation*. Monographs in Oceanographic Methodology, 4. (ed. H. F. Steedman), pp. 49–53. Unesco Press, Paris.

Beklemishev, C. W. (1971). Distribution of plankton as related to micropalaeontology. In *The micropalaeontology of oceans*. (ed. B. M. Funnel and W. R. Riedel), pp;. 75–87. Cambridge University Press.

Beklemishev, C. W. (1981). Biological structure of the Pacific Ocean as compared with two other oceans. *Journal of Plankton Research*, 3, 531–49.

Beraneck. E. (1895). Les chétognathes de la baie d'Amboine (voyage de MM. M. Bedot et C. Pictet dans l'archipel Malais). *Revue Suisse de Zoologie et Annales du Musée d'Histoire naturelle de Genève*, 3, 137–59.

Bieri, R. (1959). The distribution of the planktonic Chaetognatha in the Pacific and their relationship to the water masses. *Limnology and Oceanography*, 4, 1–28.

Bieri, R. (1966). A pale blue chaetognath from Tanabe Bay. *Publications of the Seto marine biological Laboratory*, 14, 21–2.

Bieri, R. (1974a). First record of the chaetognath genus *Krohnittella* in the Pacific and description of a new species. *Wasmann Journal of Biology*, 32, 297–301.

Bieri, R. (1974b). A new species of *Spadella* (Chaetognatha) from California. *Publications of the Seto marine biological Laboratory*, 21, 281–6.

Bieri, R. (1977). A third blue chaetognath and notes on the distribution of hyponeuston, observed *in situ*. *Publications of the Seto Marine Biological Laboratory*, 24, 27–8.

Bieri, R. (1991a). Poorly preserved chaetognaths resulting in the rejection of one family and three genera with the establishment of two new families. *Proceedings of the Biological Society of Washington*. (In press.)

Bieri, R. (1991b). A re-examination of *Bathybelos*. *American Scientist* (In press).

Bieri, R., Bonilla, D., and Arcos, F. (1983). Function of the teeth and vestibular organ in the Chaetognatha as indicated by scanning electron microscope and other observations. *Proceedings of the Biological Society of Washington*, 96, 110–14.

Biersteker, R. H., and van der Spoel, S. (1966). *Sagitta batava* n. sp. from the Scheldt estuary, the Netherlands (Chaetognatha). *Beaufortia*, 14. 61–9.

Bigelow, H. B. (1924). Plankton of the offshore waters of the Gulf of Maine. *Bulletin of the Bureau of Fisheries. Washington, DC*, 40, 1–567.

Bogorov, B. G. (1939). Weight and ecological features of the macroplankton organisms of the Barents Sea. *Transactions of the Institute of marine Fisheries and*

Oceanography USSR, **4**, 245–58. (In Russian, English summary)

Bogorov, B. G. (1940). On the biology of Euphausiidae and Chaetognatha in the Barents Sea. *Byulleten' Moskovskogo Obshchestva Istpytatelei Prorody. Moskva*, **49**, 3–18. (In Russian, English summary)

Bogorov, B. G. (1957). Unification of plankton research. *Annales Biologiques, Copenhague*, **33**, 299–315.

Bolles Lee, A. (1888). La spermatogenèse chez les chaetognathes. *Cellule*, **6**, 107–33.

Boltovskoy, D. (1975). Some biometrical, ecological, morphological and distributional aspects of Chaetognatha. *Hydrobiologia*, **46**, 515–34.

Boltovskoy, D. (1981*a*). *Atlas del zooplancton del Atlantico sudoccidental*, pp. 1–936. Publication INIDEP, Mar del Plata, Argentina.

Boltovskoy, D. (1981*b*). Chaetognatha: Morfologia. Caracteres de importancia sistemática. In *Atlas del zooplancton del Atlantico sudoccidental*, (ed. D. Boltovskoy), pp. 759–80. Publication INIDEP, Mar del Plata, Argentina.

Bone, Q. and Pulsford, A. (1978). The arrangement of ciliated sensory cells in *Spadella* (Chaetognatha). *Journal of the Marine Biological Association of the United Kingdom*, **58**, 565–70.

Bone, Q. and Pulsford, A. (1984). The sense organs and ventral ganglion of *Sagitta* (Chaetognatha). *Acta Zoologica. Stockholm*, **65**, 209–20.

Bone, Q. and Ryan, K. P. (1972). Osmolarity of osmium tetroxide and glutaraldehyde fixatives. *Histochemical Journal*, **4**, 331–47.

Bone, Q. and Ryan, K. P. (1979). The Langerhans receptor of *Oikopleura* (Tunicata: Larvacea). *Journal of the Marine Biological Association of the United Kingdom*, **59**, 69–75.

Bone, Q., Ryan, K. P., and Pulsford, A. L. (1983). The structure and composition of the teeth and grasping spines of chaetognaths. *Journal of the Marine Biological Association of the United Kingdom*, **63**, 929–39.

Bone, Q., Brownlee C., Bryan, G. W., Burt, G. R., Dando, P. R., Liddicoat, M. I., Pulsford, A. L., and Ryan, K. P. (1987*a*). On the differences between the two 'indicator' species of chaetognath, *Sagitta setosa* and *S.elegans. Journal of the Marine Biological Association of the United Kingdom*, **67**, 545–60.

Bone, Q., Grimmelikhuijzen, C. J. P., Pulsford, A., and Ryan, K. P. (1987*b*). Possible transmitter functions of acetylcholine and an RFamide-like substance in *Sagitta* (Chaetognatha). *Proceedings of the Royal Society of London B*, **230**, 1–14.

Bordas, M. (1920). Estudio de la ovogénesis de la *Sagitta bipunctata. Trabajos Museo Nacional de Ciencias Naturales. Madrid Serie Zoologia*, **42**, 5–119.

Böttger, R. and Schnack, D. (1986). On the effect of formaldehyde fixation on the dry weight of copepods. *Meeresforschung*, **31**, 141–52.

Bouligand, Y. (1985). Brisures de symétrie et morphogenèse biologique. *Compte Rendu de l'Académie La Vie des Sciences*, **2**, 121–40.

Bowman, T. E. and Bieri, R. (1990). *Paraspadella anops*, new species, from Sagittarius Cave Grand Bahama Island, the second troglobitic chaetognath. *Proceedings of the Biological Society of Washington.*

Brinton, E. (1962). The distribution of Pacific euphausiids. *Bulletin of the Scripps Institution of Oceanography*, **8**, 51–270.

Bucklin, A. and Marcus, N. H. (1985). Genetic differentiation of populations of the planktonic copepod *Labidocera aestiva. Marine Biology*, **84**, 219–24.

Burfield, S. T. (1927). *Sagitta. L.M.B.C. Memoirs Liverpool*, **28**, 1–104.

Busch, W. (1851). *Beobachtungen über Anatomie und Entwicklung einiger wirbellosen Seethiere*, pp. 1–143. A. Hirschwald, Berlin.

Bushing, M. and Feigenbaum, D. (1984). Feeding by an expatriate population of *Sagitta enflata. Bulletin Marine Science*, **34**, 240–3.

Camatini, M. and Lanzavecchia, G. (1966). Osservazioni preliminari sull'ultrastruttura della musculatura striata dei chetognati. *Atti della Accademia Nazionale dei Lincei. Memorie. Roma*. ser. VIII, **41**, 392–5.

Canino, M. F. (1981). Aspects of the nutritional ecology of *Sagitta tenuis* (Chaetognatha) in the lower Chesapeake Bay. Unpublished MA Thesis. College William and Mary, USA, pp. 81.

Canino, M. F. and Grant, G. C. (1985). The feeding and diet of *Sagitta tenuis* (Chaetognatha) in the Lower Chesapeake Bay. *Journal of Plankton Research*, **7**, 175–88.

Carillo, B. G. E., Miller, C. B., and Wiebe, P. H. (1974). Failure of interbreeding between Atlantic and Pacific populations of the marine calanoid copepod *Acartia clausi* Giesbrecht. *Limnology and Oceanography*, **19**, 452–8.

Carpenter, E. J., Anderson, S. J., Harvey, G. R., Miklas, H. P., and Peck, B. B. (1972). Polystyrene spherules in coastal waters. *Science*, **178**, 749–50.

Casanova, J-P. (1977). La fauna pélagique profonde

(Zooplancton et micronecton) de la province Atlanto-Méditerranéenne, pp. 455. Unpublished PhD Thesis. Université de Provence, Marseille, France.

Casanova, J-P. (1985*a*). Les chaetognathes de la Mer Rouge: remarques morphologiques et biogégraphiques, description de *Sagitta erythraea* sp. n. *Rapport et Procès-verbaux des Réunions de la Commission internationale pour l'Exploration scientifique de la Mer Méditerranée*, **29**, 269–74.

Casanova, J-P. (1985*b*). *Sagitta lucida* et *Sagitta adenensis*, Chaetognathes mésoplanctoniques nouveaux du nord-ouest de l'Océan Indien. *Revue des Travaux de l'Institut des Pêches maritimes*, **47**, 25–35.

Casanova, J-P. (1985*c*). Description de l'appareil génital primitif du genre *Eukrohnia* et nouvelle classification des chaetognathes. *Compte rendu de l'Académie des Sciences. Paris*, **301** ser.III, 397–402.

Casanova, J-P. (1986*a*). Quatre nouveaux Chaetognathes atlantiques abyssaux (Genre *Heterokrohnia*): description, remarques éthologiques et biogéographiques. *Oceanologia Acta*, **9**, 469–77.

Casanova, J-P. (1986*b*). *Archeterokrohnia rubra* n. gen., n. sp., nouveau Chaetognathe abyssal de l'Atlantique nord-africain: description et position systématique, hypothèse phylogénétique. *Bulletin du Muséum National d'Histoire Naturelle. Paris*, 4 sér., **8**, 185–94.

Casanova, J-P. (1986*c*.). Deux nouvelles espèces d'*Eukrohnia* (Chaetognathes) de l'Atlantique sud-tropical africain. *Bulletin du Muséum National d'Histoire Naturelle. Paris*, 4 sér., **8**, 819–33.

Casanova, J-P. (1986*d*). Découverte en Méditerranée d'un chaetognathe nouveau du genre archaique profond *Archeterokrohnia*: description et signification biogéographique. *Rapport et Procès-verbaux des Réunions de la Commission internationale pour l'Exploration scientifique de la Mer Méditerranée*, **30**, 196.

Casanova, J-P. (1986*e*). *Spadella ledoyeri*, Chaetognathe nouveau de la grotte sous-marine obscure des Tremies (Calanques de Cassis) *Rapport et Procès-verbaux des Réunions de la Commission internationale pour l'exploration scientifique de la Mer Méditerranée*, **30**, P-1115, 196.

Casanova, J-P. (1987). Deux Chaetognathes benthiques nouveaux du genre *Spadella* de parages de Gibraltar. Remarques phylogénétiques. *Bulletin du Muséum National d'Histoire Naturelle. Paris*, 4 sér., **9**, 375–90.

Casanova, J-P. (1990). A new species of *Paraspadella*

(Chaetognatha) from the coastal waters of Japan. *Proceedings of the Biological Society of Washington*, **103**, 773–83.

Casanova, J-P. and Chidgey, K. C. (1987). Une nouvelle espèce d'*Heterokrohnia* (Chaetognathe) des campagnes du 'Discovery' dans l'Atlantique nord-oriental. *Bulletin du Muséum National d'Histoire Naturelle. Paris*, 4 sér. **9**, 879–85.

Casanova, J-P. and Chidgey, K. C. (1990). A new benthopelagic species of *Heterokrohnia* (Chaetognatha) from the North Atlantic Ocean. *Bulletin. Zoologisch Museum, Universiteit van Amsterdam*, **12**, 109–16.

Casanova, J-P. and Andreu, P. (1989, publ. 1990). Les chaetognathes des pêches profondes du 'Magga Dan' le long des côtes sud et est-africaines. *Indo-Malayan Zoology*, **6**, 207–21.

Cheney, J. (1985). Spatial and temporal abundance patterns of oceanic chaetognaths in the western North Atlantic I. Hydrographic and seasonal abundance patterns. *Deep-Sea Research*, **32**, 1041–59.

Childress, J. J. and Nygaard, M. (1974). Chemical composition and buoyancy of midwater crustaceans as function of depth of occurrence off southern California. *Marine Biology*, **27**, 225–38.

Claparède, A. R. E. (1863). *Boebachtungen über Anatomie und Entwicklungsgeschichte wirbellosen Thiere an der Küste der Normandie angestellt*, pp. 1–120. W. Engelmann, Leipzig.

Clarke, G. L. and Zinn, D. J. (1937). Seasonal production of zooplankton off Woods Hole with special reference to *Calanus finmarchicus*. *Biological Bulletin. Marine Biological Laboratory, Woods Hole*, **73**, 464–87.

Clarke, G. L., Pierce, E. L., and Bumpus, D. F. (1943). The distribution and reproduction of *Sagitta elegans* on Georges Bank in relation to the hydrographical conditions. *Biological Bulletin. Marine Biological Laboratory, Woods Hole*, **85**, 201–26.

Conant, F. S. (1896). Notes on the chaetognaths. *Annals and Magazine of Natural History, London*, **18**, 201–13.

Conover, R. J. (1978). Transformation of organic matter. In *Marine Ecology* 4 (ed. O. Kinne) pp. 221–499. Wiley, New York.

Conover, R. J. and Paranjape, M. A. (1977). Comments on the use of a deep tank in planktological research. *Helgoländer wissenschaftliche Meeresuntersuchungen*, **30**, 105–17.

Conway, D. V. P. and Williams, R. (1986). Seasonal

population structure, vertical distribution and migration of the chaetognath *Sagitta elegans* in the Celtic Sea. *Marine Biology*, **93**, 377–87.

Conway-Morris, S. (1977). A description of the Middle Cambrian worm *Amiskwia sagittiformis Walcott* from the Burgess Shale of British Colombia. *Palaeontologische Zeitschrift*, **51**, 271–87.

Conway-Morris, S. (1987). The search for the Precambrian-Cambrian boundary. *American Scientist*, **75**, 156–67.

Cosper, T. C. and Reeve, M. R. (1970). Structural details of the mouthparts of a chaetognath, as revealed by scanning elecron microscopy. *Bulletin Marine Science*, **20**, 441–5.

Cosper, T. C. and Reeve, M. R. (1975). Digestive efficiency of the chaetognath *Sagitta hispida* Conant. *Journal of Experimental Marine Biology and Ecology*, **17**, 33–8.

Croce, N. D. (1963). Osservazioni sull'alimentazione di *Sagitta. Rapport et Procès-verbaux des Réunions de la Commission internationale pour l'Exploration scientifique de la Mer Méditerranée*, **17**, 627–30.

d'Orbigny, A. (1843). *Voyage dans l'Atlantique meridionale, executé dans le cours des années 1826–1833, Mollusques*, vol. 5, I–XLIII, 1–758, Paris, Strasbourg, 1835–1844.

Dallot, S. (1968). Observations préliminaires sur la reproduction en élevage du chaetognathe planctonique *Sagitta setosa* Müller. *Rapports et Procès-verbaux des Réunions de la Commission internationale pour l'Exploration scientifique de la Mer Méditerranée*, **19**, 521–3.

Dallot, S. (1970). L'Anatomie du tube digestif dans la phylogénie et la systematique des chaetognathes. *Bulletin du Muséum National d'Histoire Naturelle. Paris*, **42**, 549–65.

Dallot, S. (1971). Les chaetognathes de Nosy Bé: description de *Sagitta galerita* sp.n. *Bulletin. Zoologisch Museum, Universiteit van Amsterdam*, **2**, 13–8.

Dallot, S. and Ducret, F. (1969). Un chaetognathe mesoplanctonique nouveau *Sagitta megalophthalma* sp.n.. *Beaufortia*, **17**, 13–20.

Dallot, S. and Ibanez, F. (1972). Etude préliminaire de la morphologie et de l'évolution chez les Chaetognathes. *Investigacion Pesquera. Barcelona*, **36**, 31–41.

Dallot, S. and Laval, P. (1974.) Les chaetognathes de Nosy Bé: *Sagitta littoralis* sp.n.. *Cahier de l'Office de la Recherche Scientifique et Technique Outre-Mer Série océanographique*, **12**, 878–97.

Darwin, C. (1844). Observations on the structure and propagation of the genus *Sagitta. Annals and Magazine of Natural History, London*, **13**, 1–6.

David, P. M. (1955). The distribution of *Sagitta gazellae* Ritter-Záhony. *Discovery Reports*, **27**, 235–78.

David, P. M. (1956). *Sagitta planctonis* and related forms. *Bulletin of the British Museum (Natural History) Zoology*, **4**, 437–51.

David, P. M. (1958a). The distribution of the Chaetognatha of the southern ocean. *Discovery Reports*, **29**, 199–228.

David, P. M. (1958b). A new species of *Eukrohnia* from the southern ocean with a note on fertilization. *Proceedings of the Zoological Society of London*, **131**, 597–606.

David, P. M. (1961). The influence of vertical migration on speciation in the oceanic zooplankton. *Systematic Zoology*, **10**, 10–16.

David, P. M. (1963). Some aspects of speciation in the Chaetognatha. *Publications. Systematics Association. Speciation in the Sea*, **5**, 129–43.

Dawes, B. (1958). *Sagitta* as host of larval trematodes, including new and unique type of cercaria. *Nature. London*, **182**, 960–1.

Dawes, B. (1959). On *Cercaria owreae* (Hutton, 1954) from *Sagitta hexaptera* (d'Orbigny) *in the Caribbean. Journal of Helminthology*, **33**, 209–22,.

Dawson, J. K. (1968). Chaetognaths from the Arctic Basin, including the description of a new species of *Heterokrohnia. Bulletin of the Southern California Academy of Sciences*, **67**, 112–24.

Deevey, G. B. (1952). A survey of the zooplankton of Block Island Sound, 1943–6. *Bulletin Bingham Oceanographic Collections*, **13**, 65–119.

Deevey, G. B. (1956). Oceanography of Long Island Sound, 1952–4. *Bulletin Bingham Oceanographic Collections*, **15**, 113–55.

Deurs, B. van (1972). On the ultrastructure of the mature spermatozoon of a chaetognath, *Spadella cephaloptera. Acta Zoologica*, **53**, 93–104.

Dollfus, R. P. (1960). Distomes des Chaetognathes. *Bulletin de l'Institut des Pêches maritimes Marocain*, **4**, 19–45.

Dollfus, R. P., Anantaraman M., and Nair R. V. (1954). Métacercaire d'accocoeliidé chez *Sagitta inflata* Grassi et larve de tétraphyllide fixée à cette métacercaire. *Annales de Parasitologie humaine et comparée*, **29**, 521–526.

Doncaster, L. (1902). On the development of *Sagitta*: with notes on the anatomy of the adult. *Quarterly Journal Microscopical Science*, **46**, 351–98.

Dress, F. and Duvert, M. (1983). Etude sterologique

de la croissance des fibres de la musculature primaire de *Sagitta setosa* (Chaetognathe). *Biol. Cell.*, **48**, 2a.

Drits, A. V. (1981). Some patterns of feeding of *Sagitta enflata*. *Oceanology*, **21**, 624–8.

Ducret, F. (1965). Les espèces du genre *Eukrohnia* dans les eaux équatoriales et tropicales africaines. *Cahiers de l'Office de la Recherche Scientifique et Technique Outre-Mer (Océanographie)*, **3**, 63–78.

Ducret, F. (1968). Chaetognathes des campagnes de l''Ombango' dans les eaux équatoriales et tropicales africaines. *Cashiers de l'Office de la Recherche Scientifique et Technique Outre-Mer (Océanographie)*, **6**, 95–141.

Ducret, F. (1973). Contribution à l'étude des Chaetognathes de la mer Rouge. *Beaufortia*, **20**, 135–53.

Ducret, F. (1975). Structure et ultrastructure de l'oeil chez les Chaetognathes (genres *Sagitta* et *Eukrohnia*). *Cahiers de Biologie Marine*, **16**, 287–300.

Ducret, F. (1977). Structure et ultrastructure de l'oeil chez les Chaetognathes (genres *Sagitta* et *Eukrohnia*), pp. 119. Unpublished PhD Thesis, Universite de Provence, Marseille, France.

Ducret, F. (1978). Particularités structurales de système optique chez deux chaetognathes (*Sagitta tasmanica* et *Eukrohnia hamata*) et incidences phylogénétiques. *Zoomorphologie*, **91**, 201–15.

Dunbar, M. J. (1940). On the size distribution and breeding cycles of four marine planktonic animals from the Arctic. *Journal of Animal Ecology*, **9**, 215–26.

Dunbar, M. J. (1941). The breeding cycle in *Sagitta elegans arctica* Aurivillius. *Canadian Journal of Research*, **19**, 258–66.

Dunbar, M. J. (1962). The life cycle of *Sagitta elegans* in arctic and subarctic seas, and the modifying effects of hydrographic differences in the environment. *Canadian Journal of Marine Ecology*, **20**, 76–91.

Duvert, M. (1969*a*). Ultrastructure des myofibrilles dans les muscles longitudinaux du tronc de *Sagitta setosa* (Chaetognathe). *Compte rendu de l'Académie des Sciences. Paris*, **268**, 2452–4.

Duvert, M. (1969*b*). Sur l'existence de fibres musculaires particulières dans les muscles longitudinaux du tronc de *Sagitta setosa*. *Compte rendu de l'Académie des Sciences. Paris*, **268**, 2707–9.

Duvert, M. and Barets, A. L. (1983). Ultrastructural studies of neuromuscular junctions in visceral and skeletal muscles of the chaetognath *Sagitta setosa*. *Cell and Tissue Research*, **233**, 657–69.

Duvert, M. and Gros, D. (1982). Further studies on the junctional complex in the intestine of *Sagitta*

setosa. Freeze fracture of the pleated septate junction. *Cell and Tissue Research*, **255**, 663–71.

Duvert, M. and Salat, C. (1979). Fine structure of muscle and other components of the trunk of *Sagitta setosa* (Chaetognatha). *Tissue and Cell*, **11**, 217–30.

Duvert, M. and Salat, C. (1980). The primary body-wall musculature in the arrowworm *Sagitta setosa* (Chaetognatha): an ultrastructural study. *Tissue and Cell*, **12**, 723–38.

Duvert, M. and Salat, C. (1991). Ultrastructural studies on the fins of chaetognaths. *Tissue and Cell*, **22**, 853–63.

Duvert, M. and Savineau, J. P. (1986). Ultrastructural and physiological studies of the contraction of the trunk musculature of *Sagitta setosa* (Chaetognatha). *Tissue and Cell*, **18**, 937–52.

Duvert, M., Gros, D., and Salat, C. (1980*a*). The junctional complex in the intestine of *Sagitta setosa* (Chaetognatha): the paired septate junction. *Journal of Cell Science*, **42**, 227–46.

Duvert, M., Gros, D., and Salat, C. (1980*b*). Ultrastructural studies of the junctional complex in the musculature of the arrow-worm *Sagitta setosa* (Chaetognatha). *Tissue and Cell*, **12**, 1–11.

Duvert, M., Bouligand, Y., and Salat, C. (1984). The liquid crystalline nature of the cytoskeleton in certain epidermal cells of *Sagitta setosa*. *Tissue and Cell*, **16**, 469–81.

Duvert, M., Dress, F., and Salat, C. (1987). Etude quantitative et qualitative de la croissance du muscle du tronc d'un animal pourvu d'un hydrosquelette: *Sagitta setosa* (Chaetognathe). *Biologie Cellulaire.*, **60**, 33a.

Duvert, M., Grandier-Vazeille, X., and Chevallier, J. P. (1988*a*). Cytoenzymological and biochemical studies on the locomotor muscle of *Sagitta* (Chaetognath); comparisons with the visceral muscle. In *Sarcomeric and non-sarcomeric muscles: basic and applied research prospects for the 90's* (ed. U. Carraro). pp. 567–72, Unipress, Padua.

Duvert, M., Grandier-Vazeille, X., and Chevallier, J. P. (1988*b*). Cytochemical detection of calcium ATPase in two kinds of fibers in the locomotor muscle of *Sagitta setosa* (Chaetognatha). *Biol. Cell.*, **63**, 13A.

Eakin, R. M. and Westfall, J. A. (1964). Fine structure of the eye of a chaetognath. *Journal of Cell Biology*, **21**, 115–31.

Elian, L. (1960). Observations systématiques et biologiques sur les chaetognathes qui se trouvent dans les eaux roumaines de la Mer noire. *Rapport et Procès-verbaux des Réunions de la Commission internationale pour l'Exploration scientifique de la Mer Méditerranée*, **15**, 359–66.

Engstrøm, A. and Strandberg, B., ed. (1968). Symmetry and function of biological systems at the micromolecular level. *Proceedings of the 11th Nobel Symposium, Lidingoe*. Almqvist and Wiksell, Stolkholm.

Esterly, C. O. (1919). Reactions of various planktonic animals with reference to their diurnal migrations. *University of California Publications in Zoology*, **19**, 1–83.

Evans, M. H. (1972). Tetrodotoxin, saxotoxin, and related substances: their applications in neurobiology. *International Review of Neurobiology*, **15**, 83–166.

Fagetti, E. G. (1958). Quetognato neuvo procedente del archipiélago de Juan Fernández. *Revista de Biologia Marina. Valparaiso*, **8**, 125–31.

Fagetti. E. G. (1968). Quetognatos de la expedición 'Marchile I' con observaciones acerca del posible valor de algunas especies como indicadoras de las masas de agua frente a Chile. *Revista de Biologia Marina. Valparaiso*, **13**, 85–171.

Fasham, M. and Angel, M. V. (1975). The relationship of the zoogeographic distributions of the planktonic ostracods in the north-east Atlantic to the water masses. *Journal of the Marine Biological Association of the United Kingdom*, **55**, 739–57.

Feigenbaum, D. L. (1976). Development of the adhesive organ in *Spadella schizoptera* (Chaetognatha) with comments on growth and pigmentation. *Bulletin of Marine Science*, **26**, 600–3.

Feigenbaum, D. L. (1977). Nutritional ecology of the Chaetognatha with particular reference to external hair patterns. Ph.D. Thesis, University of Miami, Florida. pp. 106.

Feigenbaum, D. L. (1978). Hair fan patterns in the Chaetognatha. *Canadian Journal of Zoology*, **56**, 536–46.

Feigenbaum, D. L. (1979a). Daily ration of the chaetognath *Sagitta enflata*. *Marine Biology*, **54**, 78–82.

Feigenbaum, D. L. (1979b). Predation on chaetognaths by typhloscolecid polychaetes: one explanation for headless chaetognaths. *Journal of the Marine Biological Association of the United Kingdom*, **59**, 631–3.

Feigenbaum, D. L. (1982). Feeding of the chaetognath *Sagitta elegans*, at low temperatures at Vineyard Sound, Massachusetts. *Limnology and Oceanography*, **27**, 699–706.

Feigenbaum, D. L. and Maris, R. C. (1984). Feeding in the Chaetognatha. *Annual Review of Oceanography and Marine Biology*, **22**, 343–92.

Feigenbaum, D. L. and Reeve, M. R. (1977). Prey detection in the Chaetognatha: response to a vibrating probe and experimental determination of attack distance in large aquaria. *Limnology and Oceanography*, **22**, 1052–8.

Fleminger, A. (1979). *Labidocera* (Copepoda, Calanoida), new and poorly known Caribbean species with a key to species in the western Atlantic. *Bulletin of Marine Science*, **29**, 170–90.

Fleminger, A. and Hulsemann, K. (1973). Relationship of Indian Ocean epiplanktonic calanoids to the world oceans. In *The biology of the Indian Ocean* (ed. B. Zeitschel and S. A. Gerlach), pp. 339–47. Springer Verlag, Heidelberg.

Fleminger, A., Othman, B. H. R., and Greenwood, J. G. (1982). The *Labidocera pectinata* group: an Indo-Pacific lineage of planktonic copepods with descriptions of two new species. *Journal of Plankton Research*, **4**, 245–70.

Fraser, J. H. (1952). The Chaetognatha and other zooplankton of the Scottish area and their value as biological indicators of hydrographic conditions. *Marine Research Scotland*, **2**, 5–52.

Frost, B. W. and Fleminger, A. (1968). A revision of the genus *Clausocalanus* (Copepoda; Calanoida) with remarks on distributional patterns in diagnostic characters. *Bulletin of the Scripps Institution of Oceanography*, **12**, 1–235.

Fulton, R. S., III (1984). Effects of chaetognath predation on nutrient enrichment on enclosed estuarine copepod communities. *Oecologia (Berlin)*, 97–101.

Fuhrman, F. A. (1986). Tetrodotoxin, tarichatoxin, and chiriquitoxin: historical perspectives. *Annals of the New York Acadamy of Science*, **479**, 1–14.

Furnestin, M.-L. (1953). Chaetognathes récoltés en Méditerranée par le 'Président Théodore Tissier'. *Bulletin de la Station d'Aquiculture et de Pêche de Castiglione*, **4**, 275–317.

Furnestin, M.-L. (1957). Chaetognathes et zooplancton du secteur Atlantique Marocain. *Revue des Travaux de l'Institut des Pêches Maritimes*, **21**, 1–356.

Furnestin, M.-L. (1961). Compléments à l'étude de *Sagitta euxina* variété de *Sagitta setosa*. *Rapports et Procès-verbaux des Réunions de la Commission internationale pour l'Exploration scientifique de la Mer Méditerranée*, **16**, 97–101.

Furnestin, M.-L. (1965). Variations morphologiques des crochets au cours du développement dans le genre *Eukrohnia*. *Revue des Travaux de l'Institut des Pêches maritimes*, **29**, 275–84.

Furnestin, M.-L. (1967). Contribution à l'étude histologique des Chaetognathes. *Revue des Travaux de l'Institut des Pêches maritimes*, **31**, 383–92.

Furnestin, M.-L. (1976). Fixation and preservation of Chaetognatha. In *Zooplankton fixation and preservation*. Monographs on oceanographical methodology, 4. (ed. H. F. Steedman), pp. 272–8. Unesco Press, Paris.

Furnestin, M.-L. (1977). Les dents de chaetognathes au microscope électronique à balayage. *Rapport et Procès-verbaux des Réunions de la Commission internationale pour l'Exploration scientifique de la Mer Méditerranée*, **24**, 141–2.

Furnestin, M.-L. (1979). Aspects of the zoogeography of the Mediterranean plankton. in *Zoogeography and diversity of plankton* (ed. S. van der Spoel and A. C. Pierrot-Bults), pp. 191–253. Bunge, Utrecht.

Furnestin, M.-L. (1982). Dents et organe vestibulaire des chaetognathes au microscope électronique à balayage. *Revue de Zoologie Africaine, Bruxelles*, **96**, 138–73.

Furnestin, M.-L. and Ducret, F. (1965). *Eukrohnia proboscidia*, nouvelle espèce de Chaetognathe. *Revue des Travaux de l'Institut des Pêches Maritimes*, **29**, 271–3.

Germain, L. and Joubin, L. (1916). Chétognathes provenant des campagnes des yachts Hirondelle et Princesse Alice (1885–1910). *Résultats des Campagnes Scientifiques Monaco*, **49**, 1–118.

Ghirardelli, E. (1948). Chetognati raccolti nel Mar Rosso e nell Oceano Indiano dalla nave 'Cherso'. *Bollettino di Pesca, di Piscicoltura e di Idrobiologica*, **2**, 252–70.

Ghirardelli, E. (1950). Morfologia dell'apparechio digerente nel *Sagitta minima* Grassi. *Bolletino di Zoologia della Universita di Bologna*, **17**, 555–67.

Ghirardelli, E. (1951). Cycli di maturità sessuale nelle gonadi di *Sagitta inflata* Grassi del Golfo di Napoli. *Bollettina di Zoologia*, **18**, 146–62.

Ghirardelli, E. (1952). Osservazioni biologiche e sistematiche sui Chetognati del Golfo di Napoli. *Pubblicazioni della Stazione zoologica di Napoli*, **23**, 296–312.

Ghirardelli, E. (1959*a*). L'apparato riproduttore femminile nei Chetognati. *Rendiconti. Accademia nazionale dei XL. Roma*, **10**, 1–46.

Ghirardelli, E. (1959*b*). Habitat e biologia della riproduzione nei chetognati. *Archivio di Oceanografia e Limnologia. Roma*, **11**, 11–8.

Ghirardelli, E. (1968). Some aspects of the biology of chaetognaths. *Advances in Marine Biology*, **6**, 271–375.

Gibbs, R. H. (1986). The stomioid fish genus *Eustomias* and the oceanic species concept. In *Pelagic Biogeography* (ed. A. C. Pierrot-Bults, S. van der Spoel, B. J. Zahuranec, and R. K. Johnson), pp. 98–103. Unesco Technical Papers in Marine Science, **49**.

Giguère, L. A., St-Pierre, J.-F., Bernier, B., Vézina, A., and Rondeau, J.-G. (1989). Can we estimate the true weight of zooplankton samples after chemical preservation? *Canadian Journal of Fisheries and Aquatic Sciences*, **46**, 522–7.

Gordon, A. L. (1986). Interocean exchange of thermocline water. *Journal of Geophysical Research*, **91**, 5037–46.

Gorsky, G., Dallot, S., Sardou, J., Fenaux, R., Carré, C. and Palazzoli, I. (1988). C and N composition of some northwestern Mediterranean zooplankton and micronekton species. *Journal of Experimental Marine Biology and Ecology*, **124**, 134–44.

Goto, T. and Yoshida, M. (1981). Oriented light reactions of the arrow worm, *Sagitta crassa* Tokioka. *Biological Bulletin. Marine Biological Laboratory, Woods Hole*. **160**, 419–30.

Goto, T. and Yoshida, M. (1983). The role of the eye and C.N.S. components in phototaxis of the arrow worm *Sagitta crassa* Tokioka. *Biological Bulletin. Marine Biological Laboratory, Woods Hole*. **164**, 82–92.

Goto, T. and Yoshida, M. (1984). Photoreception in Chaetognatha. In *Photoreception and vision in invertebrates* (ed. M. A. Ali), pp. 727–42. NATO ASI series. Plenum Publishing Corporation, New York.

Goto, T. and Yoshida, M. (1985). The mating sequence of the benthic arrowworm *Spadella schizoptera*. *Biological Bulletin. Marine Biological Laboratory, Woods Hole*. **169**, 328–33.

Goto, T. and Yoshida, M. (1987). Nervous system in Chaetognatha. In *Nervous systems in invertebrates* (ed. M. A. Ali), pp. 461–81. NATO ASI series A, Plenum Press, New York and London.

Goto, T. and Yoshida, M. (1988). Histochemical demonstration of a rhodopsin-like substance in the eye of the arrow worm *Spadella schizoptera* (Chaetognatha). *Experimental Biology*, **48**, 1–4.

Goto, T., Takasu, N., and Yoshida, M. (1984). A unique photoreceptive structure in the arrow worms *Sagitta crassa* and *Spadella schizoptera* (Chaetognatha). *Cell and Tissue Research*, **235**, 471–8.

Goto, T., Terazaki, M., and Yoshida, M. (1989). Comparative morphology of the eyes of *Sagitta*

(Chaetognatha) in relation to depth of habitat. *Experimental Biology*, **48**, 95–105.

Gourret, P. (1884). Considération sur la faune pélagique du Golfe de Marseille, suivées d'une étude anatomique et zoologique de la *Spadella marioni*, espèce nouvelle de l'ordre de chaetognathes. (Leuckart). *Annales du Musée d'Histoire Naturelle de Marseille*, Zoologie, **2**, Mémoirs 2, 1–175.

Grandier-Vazeille, X. and Duvert, M. (1989). Biochemical studies of the trunk musculature of *Sagitta*. *Oceanis*, **15**, 381–9.

Grandier-Vazeille, X., Duvert, M., and Chevallier, J. (1989). The use of auto-anti-idiotypes for the visualisation of acetylcholine receptors in an invertebrate skeletal muscle. *Neuroscience Letters*, **99**, 30–4.

Grassi, G. B. (1881). Anatomia comparata. Intorno ai Chetognati. Nota preliminare. *Rendiconti Reale Istituto Lombardo di Scienze e Lettere*, **14**, 199–214.

Grassi, G. B. (1883). I Chetognati. *Fauna und Flora des Golfes von Neapel. Monographie*. **5**, 1–126. W. Engelman, Leipzig.

Green, C. (1981). A clarification of the two types of invertebrate pleated septate junction. *Tissue and Cell*, **13**, 173–88.

Green, C. and Bergquist, R. (1982). Phylogenetic relationships within the invertebrates in relation to the structure of septate junctions and the development of 'occluding' junctional types. *Journal Cell Science*, **53**, 279–305.

Greenbergh, M. J. and Price, D. A. (1980). Cardioregulatory peptides in molluscs. In *Peptides: integrators of cell and tissue function* (ed. F. E. Bloom), pp. 107–26. Raven Press, New York.

Grey, B. B. (1930). Chaetognatha from the Society Islands. *Proceedings of the Royal Society of Queensland*, **42**, 62–7.

Greze, V. N. (1970). The biomass and production of different trophic levels in the pelagic communities of south seas. In *Marine food chains*, (ed. J. H. Steele), pp. 458–67. Oliver and Boyd, Edinburgh.

Hagen, W. and Kapp, H. (1986). *Heterokrohnia longicaudata*, a new species of Chaetognatha from Antarctic waters. *Polar Biology*, **5**, 181–3.

Halim, Y. and Guerguess, S. K. (1973). Chaetognathes du plancton d'Alexandrie. I. Généralités. *S. friderici* R.Z.. *Rapports et Procès-verbaux des Réunions de la Commission internationale pour l'Exploration scientifique de la Mer Méditerranée*, **21**, 493–6.

Halvarson, M. and Afzelius, B. (1969). Filament organization in the body muscles of the arrow worm. *Journal of Ultrastructure Research*, **26**, 289–95.

Hamon, M. (1951). Note sur une Grégarine parasite du tube digestif de *Sagitta lyra*. *Bulletin de la Société d'Histoire Naturelle de l'Afrique de Nord*, **42**, 11–14.

Hamon, M. (1956). Chaetognathes recuellis dans la Baie de Nhatrang. *Bulletin du Muséum National d'Histoire Naturelle. Paris*, **28**, 466–73.

Hamon, M. (1957). Note sur *Janicki pigmentifera*. *Bulletin de la Société d'Histoire Naturelle de l'Afrique de Nord*. **48**, 220–33.

Harshbarger, J. C., Charles, A. M., and Spero, P. M. (1981). Collection and analysis of neoplasms in sub-homeothermic animals from a phyletic point of view. In *Phyletic approaches to cancer* (ed. C. J. Dawe, J. C. Harshbarger, S. Kondo, T. Sugimura and S. Takayama), pp. 357–84. Japan Sci. Soc. Press, Tokyo.

Hay, D. E. (1981). Effects of capture and fixation on gut contents and body size of Pacific herring larvae. *Rapports et Procès-verbaux des Réunions. Conseil permanent international pour l'Exploration de la Mer, Copenhague*, **178**, 395–400.

Hay, D. E. (1982). Fixation shrinkage of herring larvae: Effects of salinity, formalin concentration, and other factors. *Canadian Journal of Fisheries and Aquatic Sciences*, **39**, 1138–43.

Hay, D. E. (1984). Weight loss and change of condition factor during fixation of Pacific herring *Clupea harengus pallasi*, eggs and larvae. *Journal of Fish Biology*, **25**, 421–33.

Hengeveld, R. (1988). Mayr's ecological species criterion. *Systematic Biology*, **37**, 47–55.

Heron, A. C., McWilliams, P. S., and Dal Pont, G. (1988). Length-weight relation in the salp *Thalia democratica* and potential of salps as a source of food. *Marine Ecology*, **42**, 125–32.

Hertwig, O. (1880*a*). Die Chaetognathen: Eine monographie. *Jenaische Zeitschrift für Medizin und Naturwissenschaften*, **14**, 1–111.

Hertwig, O. (1880*b*). Über die Entwicklungsgeschichte der Sagitten. *Jenaische Zeitschrift für Medizin und Naturwissenschaften*, **14**, 196–303.

Hertwig, O. (1880*c*). *The Chaetognatha, their anatomy, systematics and developmental history*, pp. 86. Translated by Egan, S. M. and Sund, P. N. Library of Scripps Institution of Oceanography.

Hida, T. S. (1957). Chaetognaths and pteropods as biological indicators in the North Pacific. *United States Fish and Wildlife Service Special Scientific Report, Fisheries Series*, **215**, 1–13.

Hirota, R. (1959). On the morphological variation of

Sagitta crassa. Journal of the Oceanographical Society of Japan, **15**, 191–202.

Hirota, R. (1981). Dry weight and chemical composition of the important zooplankton in the Sentonaikai (Inland Sea of Japan). *Bulletin of Plankton Society of Japan*, **28**, 19–24.

Holmes, J. C. and Bethel, W. (1972). Modifications of intermediate host behaviour by parasites. In *Behavioural aspects of parasite transmission* (ed. E. V. Canning), pp. 123–149. Academic Press, New York.

Hopkins, T. L. (1982). The vertical distribution of zooplankton in the eastern Gulf of Mexico. *Deep-Sea Research*, **29**, 1069–83.

Horridge, G. A. and Boulton, P. S. (1967). Prey detection by Chaetognatha via a vibration sense. *Proceedings of the Royal Society of London*, **168**, 413–9.

Hovasse, R. (1924). *Trypanoplasma sagittae* sp. nov. *Compte Rendu des Séances de la Société de Biologie. Paris*, **91**, 1254–5.

Huntsman, A. G. (1919). Some quantitative and qualitative studies of the eastern Canadian plankton. A special study of the Canadian chaetognaths, their distribution, etc., in the waters of the eastern coast. *Report of the Canadian Fisheries Expedition 1914–15*, pp. 421–85.

Huntsman, A. G. and Reid, M. E. (1921). The success of reproduction in *Sagitta elegans* in the Bay of Fundy. *Transactions of the Royal Canadian Institute*, **13**, 99–112.

Huq, A., Small, E., West, P. A., Huq, M. E., Rahman, R., and Colwell, R. R. (1983). Ecological relationships between *Vibrio cholerae* and planktonic crustacean copepods. *Applied Environmental Microbiology*, **45**, 275–83.

Hutton, R. F. (1954). *Metacercaria owreae* n. sp., an unusual trematode larva from Florida current chaetognaths. *Bulletin of Marine Science of the Gulf and Caribbean*, **4**, 104–9.

Hwang, D. F., Arakawa, O., Saito, T., Noguchi, T., Simidu, U., Tsukamoto, K., Shida, Y., and Hashimoto, K. (1989). Tetrodotoxin-producing bacteria from the blue-ringed octopus *Octopus maculosus*. *Marine Biology*, **100**, 327–32.

Hyman, L. H. (1959). The enterocoelous coelomates phylum Chaetognatha. In *The invertebrates: smaller coelomate groups*. Vol 5, Ch 16, pp. 1–71. McGraw-Hill, New York.

Ibañez, F., Ducret, F., and Dallot, S. (1974). Comparaison de classifications biométriques relatives à *Sagitta regularis, Sagitta bedfordii* et *Sagitta neglecta*.

Archives Zoologie Experimentale et Générale, **115**, 205–27.

Ikeda, I. (1917). A new astomatous ciliate, *Metaphrya sagittae*, gen. et sp. nov., found in the coelom of *Sagitta*. *Annotationes Zoologicae Japonenses*, **9**, 317–24.

Ikeda, T. and Kirkwood, R. (1989). Metabolism and elemental composition of a giant chaetognath *Sagitta gazellae* from the southern ocean. *Marine Biology*, **100**, 261–7.

Ikeda, T. and Skjoldal, H. R. (1989). Metabolism and elemental composition of zooplankton from the Barents Sea during early arctic summer. *Marine Biology*, **100**, 173–83.

Ives, J. D. (1985). The relationship between *Gonyaulax tamarensis* cell toxin levels and copepod ingestion rates. In *Toxic dinoflagellates* (ed. D. M. Anderson, A. W. White, and D. G. Baden), pp 413–8. Elsevier, New York.

Jägersten, G. (1940). Zur Kenntnis der Physiologie der Zeugung bei *Sagitta*. *Zoologiska Bidrag från Uppsala*, **18**, 397–413.

Jakobsen, T. (1971). On the biology of *Sagitta elegans* Verrill and *Sagitta setosa* J. Müller in inner Oslofjord. *Norwegian Journal of Zoology*, **19**, 201–25.

Jarling, C. and Kapp H. (1985). Infestation of Atlantic chaetognaths with helminths and ciliates. *Diseases of Aquatic Organisms*, **1**, 23–8.

Joh, H. and Uno, S. (1983). Zooplankton standing stock and their estimated production in Osaka Bay. *Bulletin of the Plankton Society of Japan*, **30**, 41–51.

John, C. C. (1933). Habitats, structure, and development of *Spadella cephaloptera*. *Quarterly Journal of Microscopical Science*, **75**, 625–96.

Johnson, M. W. and Brinton, E. (1963). Biological species, watermasses and currents. In *The Sea* (ed. M. N. Hlll) **2**, pp. 38–54. Interscience Publishers, New York.

Johnson, T. H. and Taylor, B. B. (1920). Notes on Australian Chaetognatha. *Proceedings of the Royal Society of Queensland*, **31**, 28–41.

Kado, Y. (1954). Notes on the seasonal variation of *Sagitta crassa. Annotationes Zoologicae Japonenses*, **27**, 52–5.

Kado, Y. and Hirota, R. (1957). Further studies on the seasonal variation of *Sagitta crassa. Journal of Science of the Hiroshima University*. **17**, 131–6.

Kaneko, T. and Colwell, R. R. (1975). Adsorption of *Vibrio parahaemolyticus* onto chitin and copepods. *Applied Microbiology*, **29**, 269–74.

Kankaala, P. and Johannson, S. (1986). The influence of individual variation in length-biomass regressions in three crustacean zooplankton species. *Journal of Plankton Research*, **8**, 1027–38.

Kapp, H. (1991*a*). *Archeterokrohnia*, Casanova, 1986, a junior synonym of *Heterokrohnia* Ritter-Záhony, 1911 (Chaetognatha), with a review of the species of *Heterokrohnia*. *Helgoländer Meeresuntersuchungen*, **45**, (in press.)

Kapp, H. (1991*b*). Redescription of *Heterokrohnia mirabilis* Ritter-Záhony, 1911 (Chaetognatha). *Helgoländer Meeresuntersuchungen*, **45**, (in press).

Kapp, H. and Hagen, W. (1985). Two new species of *Heterokrohnia* (Chaetognatha) from Antarctic waters. *Polar Biology*, **4**, 53–9.

Kapp, H. and Mathey, J. (1989). Secretions and structures of the head of *Sagitta setosa* (haetognatha). *Helgoländer Meersesuntersuchungen*, **43**, 13–18.

Kassatkina, A. P. (1971). New neritic species of chaetognaths from Possjet Bay in the Sea of Japan. *Research on marine fauna and flora of Possjet Bay of the Sea of Japan*. **8**, 265–94. Nauka Press, Leningrad (in Russian).

Kassatkina, A. P. (1982). *Chaetognaths in the seas of the USSR and adjacent waters*, pp. 136. Nauka Press, Leningrad (in Russian).

Katayama, T., Ikeda, N., and Harada, K (1965). Carotenoids in sea breams, *Chrysophrys major Temmick and Schlegel*. *Bulletin of the Japanese Society of Scientific Fisheries*, **31**, 947–52.

Kem, W. R. (1988). Worm toxins. In *Marine toxins and venoms* (ed. A. T. Tu), pp. 353–78. Marcel Dekker Inc, New York.

Khan, M. A. and Williamson, D. I. (1970). Seasonal changes in the distribution of Chaetognatha and other plankton in the eastern Irish Sea. *Journal of Experimental Marine Biology and Ecology*, **5**, 285–303.

Kimmerer, W. J. (1984). Selective predation and its impact on prey of *Sagitta enflata* (Chaetognatha). *Marine Ecology Progress Series*, **15**, 55–62.

King, K. R. (1979). The life history and vertical distribution of the chaetognath *Sagitta elegans* in Dabob Bay, Washington. *Journal of Plankton Research*, **1**, 153–67.

Kinne, O. (1980). Diseases of marine animals: general aspects. In *Diseases of marine animals* (ed. O. Kinne), **1**, pp 13–73. John Wiley, Chichester.

Kitou, M. (1966). A new species of *Sagitta* (Chaetognatha) collected off the Izu Peninsula. *La Mer*, **4**, 238–40.

Kogure, K., Do, H. K., Thuesen, E. V., Namba, K., Ohwada, K., and Simidu, U. (1988*a*). Accumulation of tetrodotoxin in marine sediment. *Marine Ecology Progress Series*, **45**, 303–5.

Kogure, K., Tamplin, M., Simidu, U., and Colwell, R. R. (1988*b*). A tissue culture assay for tetrodotoxin, saxotoxin and related toxins. *Toxicon*, **26**, 191–7.

Køie, M. (1975). On the morphology and life-history of *Opechona bacillaris* (Molin, 1819) Looss, 1907 (Trematoda, Leprocreadidae). *Ophelia*, **13**, 63–86.

Køie, M. (1979). On the morphology and life-history of *Derogenes varicus* (Müller, 1785) Looss, 1901 (Trematoda, Hemiuridae). *Zeitschrift für Parasitenkunde*, **59**, 67–78.

Kosobokova, K. N. (1980). Caloric value of some zooplankton representatives from the central Arctic Basin and the White Sea. *Oceanology*, **20**, 84–9.

Koszteyn, J. (1983). Morphological variability and individual development cycle of *Sagitta enflata* (Grassi) 1881 as compared with the shelf-water dynamics of north-west Africa. *Oceanologia*, **16**, 53–73.

Kotori, M (1975*a*). Newly-hatched larvae of *Sagitta elegans*. *Bulletin of Plankton Society of Japan*, **21**, 113–14.

Kotori, M. (1975*b*). Morphology of *Sagitta elegans* (Chaetognatha) in early larval stages. *Journal of the Oceanographic Society of Japan*, **31**, 139–44.

Kotori, M. (1976). The biology of Chaetognatha in the Bering Sea and the northwestern North Pacific Ocean, with emphasis on *Sagitta elegans*. *Memoirs of the Faculty of Fisheries of the Hokkaido University*, **23**, 95–183.

Kotori, M. (1979). Reproduction and life history of *Sagitta elegans* Verrill (review). *Bulletin of Plankton Society of Japan*, **26**, 25–39.

Kotori, M. and Kobayashi, T. (1979). Plankton investigations in Ishikari Bay, Hokkaido. IV. A brief description of five species of Chaetognatha with a note on their vertical distribution. *Bulletin of the Hokkaido Regional Fisheries Research Laboratory*, **44**, 39–55.

Kowalewsky, A. (1871). Entwickelungsgeschichte der *Sagitta*. *Mémoires de l'Académie Impériale des Sciences de St. Petersbourg*, **16**, 7–12.

Kramp, P. L. (1939). The Godthaab Expedition 1928: Chaetognatha. *Meddelelsev om Grönland*, **80**, 1–40.

Krohn, A. (1853). Nachträchliche Bemerkungen über den Bau der Gattung *Sagitta*, nebst der Beschreibung einiger neuen Arten. *Archive für Naturgeschichte*, **19**, 266–77.

Kuhl, W. (1938). Chaetognatha. In *Bronn's Klassen und Ordnungen des Tierreichs, Band 4, Vermes*, Abteilung **4**, Buch 2, Teil 1, pp. 1–226.

Kuhl, W. and Kuhl, G. (1965). Die Dynamik der Frühentwicklung von *Sagitta setosa*. *Helgoländer wissenschaftliche Meeresuntersuchungen*, **12**, 260–301.

Kuhlmann, D. (1976). Experimentelle Untersuchungen zum Nahrungserwerb der Chaetognathen *Sagitta setosa* J. Müller und *Sagitta elegans* Verrill unter besonderer Berücksichtigung ihres Verhaltens gegenüber Fischbrut. Diplom Arbeit University, Kiel, FRG.

Kuhlmann, D. (1977). Laboratory studies of the feeding behavior of the chaetognaths *Sagitta setosa* J. Müller and *Sagitta elegans* Verrill with special reference to fish eggs and larvae as food organisms. *Meeresforschung*, **25**, 163–71.

Kulmatycki, W. J. (1918). Bericht über die Regenerationsfähigkeit der *Spadella cephaloptera*. *Zoologische Anzeige*, **29**, 281–4.

Künne, C. (1952). Untersuchungen über das Grossplankton der Deutschen Bucht und im Nordsylter Wattenmeer. *Helgoländer wissenschaftliche Meeresuntersuchungen*, **4**, 1–54.

Kuroda, K. (1981). A new chaetognath, *Eukrohnia kitou*, n.sp., from the entrance to Tokyo Bay. *Publications of the Seto Marine Biological Laboratory*, **26**, 177–85.

Lasker. R. (1966). Feeding, growth, respiration and carbon utilization of a euphausiid crustacean. *Journal of the Fisheries Research Board of Canada*, **23**, 1291–317.

Lebour, M. V. (1917). Some parasites of *Sagitta bipunctata*. *Journal of the Marine Biological Association of the United Kingdom*, **11**, 201–6.

Lebour, M. V. (1922). The food of plankton organisms. *Journal of the Marine Bioligical Association of the United Kingdom*, **12**, 644–77.

Lebour, M. V. (1923). The food of plankton organisms, II. *Journal of the Marine Biological Association of the United Kingdom*, **13**, 70–92.

Lee, W. L. (1966). Pigmentation of the marine isopod, *Idothea montereyensis*. *Comparative Biochemistry and Physiology*, **18**, 17–36.

Lele, S. H. and Gae P. B. (1936). Common sagittae of the Bombay Harbour. *Journal of the University of Bombay*, **4**, 105–13.

Leuckart, R. and Pagenstecher, A. (1858). Untersuchungen über niedere Seethiere. *Archiv für Anatomie, Physiologie und wissenschaftliche Medizin*, **1858**, 558–613.

Longhurst, A. R. (1976). Vertical migration. In *The ecology of the seas* (ed. D. H. Cushing and J. J. Walsh), pp. 116–37. Blackwell, Oxford.

Lovegrove, T. (1966). The determination of the dry weight of plankton and the effect of various factors on the values obtained. In *Some contemporary studies in marine science*. (ed. H. Barnes), pp. 429–67. George Allen and Unwin, London.

Lubny-Gertzyk, Y. A. (1953). Weight characteristics of the main forms of zooplankton in the Okhotsk and Bering Seas. *Doklady Akademii Nauk USSR*, **91**, 949–52.

McIntosh, W. C. (1890). Notes from the St. Andrews Marine Laboratory (under the Fishery Board for Scotland) No. XII. *Annals and Magazine of Natural History*, **1890**, 174–85.

McIntosh, W. C. (1927). Notes from the Gatty Marine Laboratory, St. Andrews No. L. *Annals and Magazine of Natural History*, ser. 9, **20**, 1–23.

McLaren, I. A. (1963). Effects of temperature on the growth of zooplankton, and the adaptive value of vertical migration. *Journal of the Fisheries Research Board of Canada*, **20**, 685–727.

McLaren, I. A. (1966). Adaptive significance of large size and long life of the chaetognath *Sagitta elegans* in the arctic. *Ecology*, **47**, 852–5.

McLaren, I. A. (1969). Population and production ecology of zooplankton in Ogac Lake, a landlocked fjord on Baffin Island. *Journal of the Fisheries Research Board of Canada*, **26**, 1485–559.

McLean, N. and Nielsen, C. (1989). *Oodinium jordani* n.sp., a dinoflagellate (Dinoflagellata: Oodinidae) ectoparasitic on *Sagitta elegans* (Chaetognatha). *Diseases of Aquatic Organisms*, **7**, 61–6.

McLelland, J. A. (1980). Notes on the northern Gulf of Mexico occurrence of *Sagitta friderici* Ritter-Záhony (Chaetognatha). *Gulf Research Reports*, **6**, 343–8.

McLelland, J. A. (1984). Observations on chaetognath distributions in the northeastern Gulf of Mexico during the summer of 1974. *Northeast Gulf Science*, **7**, 49–59.

McLelland, J. A. (1989*a*). *Eukrohnia calliops*, a new species of chaetognath from the northern Gulf of Mexico with notes on related species. *Proceedings of the Zoological Society of Washington*, **102**, 33–44.

McLelland, J. A. (1989*b*). An illustrated key to the chaetognaths of the northern Gulf of Mexico with notes on their distribution. *Gulf Research Report*, **8**, 145–72.

Malakhov, V. V. and Frid, M. G. (1984). Structure of

the ciliary loop and retrocerebral organ in *Sagitta glacialis* (Chaetognatha). *Doklady Akademii Nauk. SSSR*, **277**, 763–5.

Martens, P. (1976). Die planktischen Sekundär- und Tertiärproduzenten im Flachwasserökosystem der Westlichen Ostsee. *Kieler Meeresforschung*, **3**, 60–71.

Marumo, R. and Kitou, M. (1966). A new species of *Heterokrohnia* (Chaetognatha) from the western North Pacific. *La Mer*, **4**, 178–83.

Marumo, R. and Nagasawa, S. (1973). Pelagic chaetognaths in Sagami Bay and Suruga Bay, Central Japan. *Journal of the Oceanographical Society of Japan*, **29**, 267–75.

Massuti Oliver, M. (1954). Sobre la biología de las *Sagitta* del plancton del Levante español. *Publicaciones del Instituto de Biologia Aplicada Barcelona*, **16**, 137–48.

Massuti Oliver, M. (1958). Estudio del crecimiento relativo de *Sagitta enflata* Grassi del plancton de Castellón. *Investigacion Pesquera. Barcelona*, **13**, 37–48.

Matthews, J. B. L. and Hestad, L. (1977). Ecological studies on the deep-water pelagic community of Korsfjorden, western Norway: length/weight relationships for some macroplanktonic organisms. *Sarsia*, **63**, 57–63.

Mawson, P. M. (1944). Some species of the chaetognath genus *Spadella* from New South Wales. *Transactions of the Royal Society of South Australia*, **68**, 327–33.

Mayr, E. (1988). *Towards a new philosophy of biology*, pp. 564. Bellknap Press, Cambridge, MA.

Meek, A. (1928). On *Sagitta elegans* and *Sagitta setosa* from the Northumbrian plankton with a note on a trematode parasite. *Proceedings of the Zoological Society of London*, **1928**, 743–76.

Michael, E. L. (1911). Classification and vertical distribution of the Chaetognatha of the San Diego Region. *University of California Publications in Zoology*, **8**, 121–86.

Michael, E. L. (1919). Report on the Chaetognatha of the Albatross expedition to the Phillipines. *Bulletin of the United States National Museum*, **100**, 235–77.

Michel, H. B. (1984). *Chaetognatha of the Caribbean Sea and Adjacent Areas*, pp. 33. National Marine Fisheries Service (US), NOAA Technical Report NMFS 15.

Miller, J. K. (1966). Biomass determination of selected zooplankters found in the California Cooperative Oceanic Fisheries Investigations. *SIO Reference*

66–15, *Marine Life Research, Scripps Institution Oceanography, San Diego*, pp. 16.

Mironov, G. N. (1960). The food of plankton predators. 2. Food of *Sagitta. Trudy Sevastopol'skoi Biologicheskoi Stantsii*, **13**, 78–88.

Moltschanoff, L. A. (1909). Die Chaetognathen des Schwarzen Meeres. *Izvestiya Imperatorskoi Akademii Nauk. S. Petersburg* (ser.6), **3**, 887–902.

Moreno, I. (1975). Estudio de la ultraestructura de la membrana basal de *Sagitta bipunctata (Quetognatos)*. *Primer Centenario de la Real Sociedad Española de Historia Natural*, 377–84.

Moreno, I. (1979). Study of the grasping spines and teeth of 6 chaetognath species observed by scanning electron microscopy. *Anatomischer Anzeiger*, **145**, 453–63.

Mosher, H. S. (1986). The chemistry of tetrodotoxin. *Annals of the New York Academy of Sciences*, **479**, 32–43.

Müller, J. (1847). Fortsetzung des Berichts über einige neue Thierformen der Nordsee. *Archiv für Anatomie, Physiologie and wissenschaftliche Medizin*, **1847**, pp. 247.

Murakami, A. (1957). The occurrence of planktonic chaetognaths in the Bay and Inland Sea regions. 1. On the occurring conditions in Tokyo Bay and the central and western parts of the Seto Inland Sea. *Suisangaku-Syusei, University of Tokyo*, **10**, 357–84 (in Japanese).

Murakami, A. (1959). Marine biological study on the planktonic chaetognaths in the Seto Inland Sea. *Bulletin Naikai Regional Fishery Research Laboratory*, **12**, 1–186.

Murakami, A. (1966). Rearing experiments of a chaetognath, *Sagitta crassa. Information Bulletin of Planktology, Japan*, **13**, 62–5.

Nagasawa, S. (1984). Laboratory feeding and egg production in the chaetognath *Sagitta crassa* Tokioka. *Journal of Experimental Marine Biology and Ecology*, **76**, 51–65.

Nagasawa, S. (1985*a*). Ecological significance of deformed chaetognaths associated with bacteria. *Bulletin of Marine Science*, **37**, 707–15.

Nagasawa, S. (1985*b*). Copulation in the neritic chaetognath *Sagitta crassa. Journal of Plankton Research*, **7**, 927–35.

Nagasawa, S. (1985*c*). The digestive efficiency of the chaetognath *Sagitta crassa* Tokioka, with observations on the feeding process. *Journal of Experimental Marine Biology and Ecology*, **87**, 271–82.

Nagasawa, S. (1985*d*). Tumour-like swellings in the

chaetognath *Sagitta crassa. Bulletin of the Plankton Society of Japan*, **32**, 75–7.

Nagasawa, S. (1986*a*). The bacterial adhesion to copepods in coastal waters in different parts of the world. *La Mer*, **24**, 117–24.

Nagasawa, S. (1986*b*). High incidence of copepod-bacteria associations in Tokyo Bay waters and Woods Hole waters. *La Mer*, **24**, 177–85.

Nagasawa, S. (1987*a*). Exoskeletal scars caused by bacterial attachment to copepods. *Journal of Plankton Research*, **9**, 749–53.

Nagasawa, S. (1987*b*). Sperm emission in the chaetognath *Sagitta crassa. Journal of Plankton Research*, **9**, 755–9.

Nagasawa. S. (1988). Copepod-bacteria associations in Zielony Lake, Poland. *Journal of Plankton Research*, **10**, 551–4.

Nagasawa, S. (1989*a*). Feeding habits of immature chaetognaths in Tokyo Bay. *Journal of Plankton Research*, **11**, 615–9.

Nagasawa, S. (1989*b*). Bacterial epibionts of copepods. *Science Progress, Oxford*, **73**, 169–76.

Nagasawa, S. and Marumo, R. (1972). Feeding of a pelagic chaetognath *Sagitta nagae Alvariño* in Suruga Bay, Central Japan. *Journal of the Oceanographic Society of Japan*, **28**, 181–6.

Nagasawa, S. and Marumo, R. (1973). Structure of grasping spines of six chaetognath species observed by scanning electron microscopy. *Bulletin of the Plankton Society of Japan*, **19**, 63–74.

Nagasawa, S. and Marumo, R. (1977). Seasonal variation in composition and number of epipelagic chaetognaths in Sagami Bay, Japan. *La Mer*, **15**, 185–95.

Nagasawa, S. and Marumo, R. (1978*a*). Reproduction and life history of the chaetognath *Sagitta nagae Alvariño* in Suruga Bay. *Bulletin of Plankton Society of Japan*, **25**, 67–84.

Nagasawa, S. and Marumo, R. (1978*b*). Fine structure of ciliary sense organs of a chaetognath *Sagitta nagae Alvariño* observed by scanning electron microscopy. *La Mer*, **16**, 7–17.

Nagasawa, S. and Marumo, R. (1979*a*). Parasites in chaetognaths in Suruga Bay, Japan. *La Mer*, **17**, 127–36.

Nagasawa, S. and Marumo, R. (1979*b*). Identification of chaetognaths based on the morphological characteristics of hooks. *La Mer*, **17**, 178–88.

Nagasawa, S. and Marumo, R. (1981). Chaetognaths as food of demersal fishes in the East China Sea.

Bulletin of the Seikai Regional Fisheries Research Laboratory, **56**, 1–13.

Nagasawa, S. and Marumo, R. (1984*a*). Parasitic infections of the chaetognath *Sagitta crassa* Tokioka in Tokyo Bay. *Bulletin of Plankton Society of Japan*, **31**, 75–7.

Nagasawa, S. and Marumo, R. (1984*b*). Feeding habits and copulation of the chaetonath *Sagitta crassa. La Mer*, **22**, 8–14.

Nagasawa, S. and Nemoto, T. (1984). X-diseases in the chaetognath *Sagitta crassa. Helgoländer Meeresuntersuchungen*, **37**, 139–84.

Nagasawa, S. and Nemoto, T. (1985). The decay of chaetognaths. *La Mer*, **23**, 56–63.

Nagasawa, S. and Nemoto, T. (1986). The widespread occurence of copepod-bacterial associations in coastal waters. *Syllogeus*, **58**, 379–84.

Nagasawa, S. and Terazaki, M. (1987). Bacterial epibionts of the deep-sea copepod *Calanus cristatus* Krøyer. *Oceanologica Acta*, **10**, 475–9.

Nagasawa, S. Simidu, U., and Nemoto, T. (1984). Bacterial invasion of chaetognaths under laboratory and natural conditions. *Journal of the Oceanographical Society of Japan*, **40**, 327–33.

Nagasawa, S., Simidu, U., and Nemoto, T. (1985*a*). Scanning electron microscopy investigation of bacterial colonization of the marine copepod *Acartia clausi. Marine Biology*, **87**, 61–6.

Nagasawa, S., Simidu, U., and Nemoto, T. (1985*b*). Ecological aspects of deformed chaetognaths and visual observations of their periphytes. *Marine Biology*, **87**, 67–75.

Nagasawa, S., Terazaki, M., and Nemoto, T. (1987*a*). Bacterial attachment to the epipelagic copepod *Acartia* and the bathypelagic copepod *Calanus. Proceedings of the Japan Academy*, ser. b, **63**, 33–5.

Nagashima, Y., Maruyama, T., Noguchi, T., and Hashimoto, K. (1987). Analysis of paralytic shellfish poison and tetrodotoxin by ion pairing high performance liquid chromatography. *Nippon Suisan Gakkaishi*, **53**, 819–23.

Nair, V. R. (1974). Distribution of chaetognaths along the salinity gradient in the Cochin backwater, an estuary connected to the Arabian Sea. *Journal of the Marine Biological Association of India*, **16**, 721–30.

Nair, V. R. (1976). Species diversity of chaetognaths along the equatorial region of the Indian Ocean with comments on the community structure. *Indian Journal of Marine Science*, **5**, 107–22.

Nair, V. R. (1977). Chaetognaths of the Indian Ocean. *Proceedings Symposium on Warm Water Zooplankton, Goa*, pp. 168–95. Special Publication Unesco/NIO, Goa.

Nair, V. R. (1978). Bathymetric distribution of chaetognaths in the Indian Ocean. *Indian Journal of Marine Science*, 7, 276–82.

Nair, V. R. (1986). Monsoon regime in the Indian Ocean and zooplankton variability. In *Pelagic Biogeography* (ed. A. C. Pierrot-Bults, S. van der Spoel, B. J. Zahurance and R. K. Johnson), pp. 210–213. Unesco Technical Papers in Marine Science, 49, Unesco, Paris.

Nair, V. R. and Madhupratap, M. (1984). Latitudinal range of chaetognaths and ostracods in the western tropical Indian Ocean. *Hydrobiology*, 112, 209–16.

Nair, V. R. and Rao, T. S. S. (1973). Chaetognatha of the Arabian Sea. In *The biology of the Indian Ocean* (ed. B. Zeitschel and S. A. Gerlach), pp. 293–317. Springer Verlag, Heidelberg.

Nair, S., Kita-Tsukamoto, K., and Simidu, U. (1988). Bacterial flora of healthy and abnormal chaetognaths. *Nippon Suisan Gakkaishi*, 54, 491–6.

Narita, H., Matsubara, S., Miwa, N., Akahane, S., Murakami, M., Goto, T., *et al* (1987). *Vibrio alginolyticus*, a TTX-producing bacterium isolated from the starfish *Astropecten polyacanthus*. *Nippon Suisan Gakkaishi*, 53, 617–21.

Newbury, T. K. (1972). Vibration perception by chaetognaths. *Nature, London*, 236, 459–60.

Newbury, T. K. (1978). Consumption and growth rates of chaetognaths and copepods in subtropical oceanic waters. *Pacific Science*, 32, 61–78.

Nicol, S. (1987). Some limitations on the use of the lipofuscin ageing techniquie. *Marine Biology*, 93, 609–14.

Noguchi, T., Hwang, D. F., Arakawa, O., Sugita, H., Deguchi, Y., Shida, Y., and Hashimoto, K. (1987). *Vibrio alginolyticus*, a TTX- producing bacterium, in the intestines of the fish *Fugu vermicularis vermicularis*. *Marine Biology*, 94, 625–30.

Noirot-Timothée, C. and Noirot, C. (1980). Septate and scalariform junctions in arthropods. *International Review of Cytology*, 63, 97–140.

Omori, M. (1969). Weight and chemical composition of some important oceanic zooplankton in the North Pacific Ocean. *Marine Biology*, 3, 4–10.

Omori, M. (1978). Some factors affecting on dry weight, organic weight and concentrations of carbon and nitrogen in freshly prepared and in preserved zooplankton. *International Revue der gesamten Hydrobiologie*, 63, 261–9.

Onoue, Y., Noguchi, T., and Hashimoto, K. (1984). Tetrodotoxin determination methods. In *Seafood Toxins* (ed. E. P. Ragelis), pp. 345–55. American Chemical Society, Washington DC.

Øresland, V. (1983). Abundance, breeding and temporal size of the chaetognath *Sagitta setosa* in the Kattegat. *Journal of Plankton Research*, 5, 425–39.

Øresland, V. (1985). Temporal size and maturity-stage distribution of *Sagitta elegans* and occurrence of other chaetognath species in Gullmarsfjorden, Sweden. *Sarsia*, 70, 95–101.

Øresland, V. (1986a). Parasites of the chaetognath *Sagitta setosa* in the western English Channel. *Marine Biology*, 92, 87–91.

Øresland, V. (1986b). Temporal distribution of size and maturity stages of the chaetognath *Sagitta setosa* in the western English Channel. *Marine Ecology Progress Series*, 29, 55–60.

Øresland, V. (1987). Feeding of the chaetognaths *Sagitta elegans* and *Sagitta setosa* at different seasons in Gullmarsfjorden, Sweden. *Marine Ecology Progress Series*, 39, 69–79.

Oug, E. (1977). Faunal distribution close to the sediment of a shallow marine environment. *Sarsia*, 63, 115–21.

Owre, H. B. (1960). Plankton of the Florida Current. Part VI. The Chaetognatha. *Bulletin Marine Science Gulf and Caribbean*, 10, 255–322.

Owre, H. B. (1963). The genus *Spadella* (Chaetognatha) in the western North Atlantic Ocean, with descriptions of two new species. *Bulletin of Marine Science of the Gulf and Caribbean*, 13, 378–90.

Owre, H. B. (1972). Marine biological investigations in the Bahamas. The genus *Spadella* and other Chaetognatha. *Sarsia*, 49, 49–58.

Owre, H. B. (1973). A new chaetognath genus and species with remarks on the taxonomy and distribution of others. *Bulletin Marine Science Gulf and Caribbean*, 23, 948–63.

Owre, H. B. and Bayer, F. M. (1962). The systematic position of the middle Cambrian fossil *Amiskwia* Walcott. *Journal of Paleontology*, 36, 1361–3.

Oye, P. van (1918). Untersuchungen über die Chaetognathen des Javameeres. *Contributions Faune Indes Néerlandaise*, 4, 1–61.

Oye, P. van (1931). La fécondation chez les chaetognathes. *Bulletin du Musée Royal d'Histoire Naturelle de Belgique*, 7, 1–7.

Paine, R. T. (1966). Food web complexity and species diversity. *American Naturalist*, **100**, 65–75.

Parry, D. A. (1944). Structure and function of the gut in *Spadella cephaloptera* and *Sagitta setosa*. *Journal of the Marine Biological Association of the United Kingdom*, **26**, 16–36.

Pathansali, D. (1974). Chaetognatha in the coastal waters of Peninsular Malaysia with descriptions of two new species. *Fisheries Bulletin Ministry of Agriculture and Rural Development Malaysia*, **2**, pp. 29.

Pathansali, D. and Tokioka, T. (1963), A new chaetognath, *Sagitta johorensis* n.sp., from Malaysian waters. *Publications of the Seto Marine Biological Laboratory*, **11**, 105–7.

Pearre, S. Jr. (1973). Vertical migration and feeding of *Sagitta elegans* Verrill. *Ecology*, **54**, 300–14.

Pearre, S. Jr. (1974). Ecological studies of three West-Mediterranean chaetognaths. *Investigacion Pesquera. Barcelona*, **38**, 325–69.

Pearre, S. Jr. (1976a). A seasonal study of the diets of three sympatric chaetognaths. *Investigacion Pesquera. Barcelone*, **40**, 1–16.

Pearre, S. Jr. (1976b). Gigantism and partial parasitic castration of chaetognatha infected with larval trematodes. *Journal of the Marine Biological Association of the United Kingdom*, **56**, 503–13.

Pearre, S. Jr. (1979a). Niche modification in chaetognatha infected with larval trematodes (Digenea). *Internationale Revue der gesamten Hydrobiologie*, **64**, 193–206.

Pearre, S. Jr. (1979b). Problems of detection and importance of vertical migration. *Journal of Plankton Research*, **1**, 29–44,

Pearre, S. Jr. (1980a). Feeding by Chaetognatha: The relation of prey size to predator size in several species. *Marine Ecology Progress Series*, **3**, 125–34.

Pearre, S. Jr. (1980b). The copepod width-weight relation and its utility in food chain research. *Canadian Journal of Zoology*, **58**, 1884–91.

Pearre, S. Jr. (1981). Feeding by Chaetognatha: Energy balance and importance of various components of the diet of *Sagitta elegans*. *Marine Ecology Progress Series*, **5**, 45–54.

Pearre, S. Jr. (1982). Feeding by Chaetognatha: Aspects of inter- and intra-specific predation. *Marine Ecology Progress Series*, **7**, 33–45.

Pereiro, J. A. (1972). Análisis de la corelación de caracteras en el quetognato *Sagitta enflata* Grassi. *Investigacion Pesquera. Barcelona*, **36**, 15–22.

Peterson, W. (1975). Distribution, abundance and biomass of the macrozooplankton of Kaneohe Bay, Oahu, Hawaii, 1966–71. *Hawai Institute of Marine Biology Technical Report*, **31**, pp. 122.

Petipa, T. S., Pavlova, E. V., and Moronov, G. N. (1970). The food web structure, utilization and transport of energy by trophic levels in the planktonic communities. In *Marine food chains* (ed. J. H. Steel), pp. 142–67. Oliver and Boyd, Edinburgh.

Pierce, E. L. (1941). The occurrence and breeding of *Sagitta elegans* Verrill and *Sagitta setosa* J. Müller in parts of the Irish Sea. *Journal of the Marine Biological Association of the United Kingdom*, **25**, 113–24.

Pierce, E. L. (1951). The Chaetognatha of the west coast of Florida. *Biological Bulletin. Marine Biological Laboratory, Woods Hole*, **100**, 206–28.

Pierrot-Bults, A. C. (1969). The synonymy of *Sagitta planctonis* and *Sagitta zetesios* (Chaetognatha). *Bulletin. Zoologisch Museum. Universiteit van Amsterdam*, **1**, 125–9.

Pierrot-Bults, A. C. (1974). Taxonomy and zoogeography of certain members of the *Sagitta serratodentata*-group (Chaetognatha). *Bijdragen tot de Dierkunde*, **44**, 215–34.

Pierrot-Bults, A. C. (1975a). Taxonomy and zoogeography of *Sagitta planctonis* Steinhaus, 1896 (Chaetognatha) in the Atlantic Ocean. *Beaufortia*, **23**, 27–51.

Pierrot-Bults, A. C. (1975b). Morphology and histology of the reproductive system of *Sagitta planctonis* Steinhaus, 1896 (Chaetognatha). *Bijdragen tot de Dierkunde*, **45**, 225–36.

Pierrot-Bults, A. C. (1976a). Zoogeographic patterns in Chaetognatha and some other planktonic organisms. *Bulletin. Zoologisch Museum. Universiteit van Amsterdam*, **5**, 59–72.

Pierrot-Bults, A. C. (1976b). Histology of the seminal vesicles in the *Sagitta serratodentata*-group (Chaetognatha). *Bulletin. Zoologisch Museum, Universiteit van Amsterdam*, **5**, 19–29.

Pierrot-Bults, A. C. (1979). On the synonymy of *Sagitta decipiens* Fowler, 1905 and *Sagitta neodecipiens* Tokioka, 1959, and the validity of *Sagitta sibogae* Fowler, 1906. *Bulletin. Zoologisch Museum, Universiteit van Amsterdam*, **6**, 137–43.

Pierrot-Bults, A. C. (1982). Vertical distribution of Chaetognatha in the central Northwest Atlantic near Bermuda. *Biological Oceanography*, **2**, 31–61.

Pierrot-Bults, A. C. and Van der Spoel, S. (1979). Speciation in macrozooplankton. In *Zoogeography and diversity of plankton* (ed. S. van der Spoel and A. C. Pierrot-Bults), pp. 144–67. Bunge, Utrecht.

Pierrot-Bults, A. C., Van der Spoel, S., Zahuranrec,

B. J., and Johnson, R. K. (1986). *Pelagic Biogeography*, pp. 1–295. Unesco Technical Papers in marine Science, **49**, Unesco, Paris.

Pineda-Polo, F. H. (1979). Seasonal distribution of the chaetognaths in the Bight of Panama. *Bol. Inst. Oceanogr. Univ. Oriente*, **18**, 65–88.

Piyakarnchana, T. (1965). The plankton community in the southern part of Kaneone Bay, Oahu, with special emphasis on the distribution, breeding season and population fluctuation of *Sagitta enflata* Grassi. Ph.D. Thesis. University of Hawai. pp. 193.

Platt, T., Brawn, V. M., and Urwin, B. (1969). Caloric and carbon equivalents of zooplankton biomass. *Journal of the Fisheries Research Board of Canada*, **26**, 2345–9.

Porumb, F. (1982). Production des principaux composants du zooplancton des eaux roumaines de la Mer Noire. *Certcet ri Mar. IRCM*, **15**, 59–67.

Prenant, A. (1912). Problèmes cytologiques généraux soulevés par l'étude des cellules musculaires. *Journal de l'Anatomie et de la Physiologie Normales et Pathologiques de l'Homme et des Animaux*, **3**,

Quoy, J. R. C. and Gaimard, P. (1827). Observations zoologiques faites à bord de 'l'Astrolabe' en Mai 1826, dans le detroit de Gibraltar. *Annales de Science naturelle (Zoologie)*, **10**, 5–239.

Rakusa-Suszczewski, S. (1969). The food and feeding habits of chaetognaths in the seas around the British Isles. *Pol.Arch. Hydrobiology*, **13**, 213–32.

Ramult, M. and Rose, M. (1945). Récherches sur les Chétognathes de la Baie d'Alger. *Bulletin de la Sociéte d'Histoire Naturelle de l'Afrique de Nord*, **36**, 45–71.

Rao, T. S. S. (1979). Zoogeography of the Indian Ocean. In *Zoogeography and diversity of plankton* (ed. S. van der Spoel and A. C. Pierrot-Bults), pp. 254–92. Bunge, Utrecht.

Rao, T. S. S. and Kelly, S. (1962). Studies on the Chaetognatha of the Indian Sea. VI. On the biology of *Sagitta enflata* Grassi in the waters of Lawson's Bay, Waltair. *Journal Zoological Society of India*, **14**, 219–25.

Rebecq, J. (1965). Considerations sur la place des trematodes dans le zooplancton marin. *Annales de Faculté des Sciences de l'Université de Aix Marseille*, **38**, 61–84.

Redfield, A. C. and Beale, A. (1940). Factors determining the distribution of populations of chaetognaths in the Gulf of Maine. *Biological Bulletin. Marine Biological Laboratory, Woods Hole*, **79**, 459–87.

Reeve, M. R. (1964). Feeding of zooplankton, with special reference to some experiments with *Sagitta*. *Nature. London*, **201**, 211–13.

Reeve, M. R. (1966). Observations on the biology of a chaetognath. In *Some contemporary studies in marine science* (ed. H. Barnes), pp. 613–30. George Allen and Unwin Ltd, London.

Reeve, M. R. (1970a). The biology of Chaetognatha I. Quantitative aspects of growth and egg production in *Sagitta hispida*. In *Marine food chains* (ed. J. H. Steele), pp. 168–89. Oliver and Boyd, Edinburgh.

Reeve, M. R. (1970b). Complete cycle of development of a pelagic chaetognath in culture. *Nature. London*, **227**, 381.

Reeve, M. R. (1971). Deadly arrow worm. *Sea Frontiers*, **17**, 175–83.

Reeve, M. R. (1977). The effect of laboratory conditions on the extrapolation of experimental measurements to the ecology of marine zooplankton. V. A review. *Proceedings Symposium on Warm Water Zooplankton. Goa*, pp. 528–37. Special Publication UNESCO/NIO, Goa.

Reeve, M. R. (1980). Comparative experimental studies on the feeding of chaetognaths and ctenophores. *Journal of Plankton Research*, **2**, 381–93.

Reeve, M. R. (1981). Large cod-end reservoirs as an aid to the live collection of delicate zooplankton. *Limnology and Oceanography*, **26**, 577–80.

Reeve, M. R. and Baker, L. D. (1975). Production of two planktonic carnivores (chaetognath and ctenophore) in south Florida inshore waters. *Fishery Bulletin US*, **73**, 238–48.

Reeve, M. R. and Cosper, T. C. (1975). Chaetognatha. In *Reproduction of marine invertebrates. 2. Entoprocts and lesser coelomates* (ed. A. C. Giese, and J. S. Pearse), pp. 157–84. Academic Press, New York.

Reeve, M. R. and Lester, B. (1974). The process of egg-laying in the chaetognath *Sagitta hispida*. *Biological Bulletin. Marine Biological Laboratory, Woods Hole*, **147**, 247–56.

Reeve, M. R. and Walter, M. A. (1972a). Observations and experiments on methods of fertilization in the chaetognath *Sagitta hispida*. *Biological Bulletin. Marine Biological Laboratory, Woods Hole*, **143**, 207–14.

Reeve, M. R. and Walter, M. A. (1972b). Conditions of culture, food-size selection and the effects of temperature and salinity on growth rate and generation time in *Sagitta hispida* Conant. *Journal of Experimental Marine Biology and Ecology*, **9**, 191–200.

Reeve, M. R., Raymont, J. E. G., and Raymont,

J. K. B. (1970). Seasonal biochemical composition and energy sources of *Sagitta hispida*. *Marine Biology*, **6**, 357–64.

Reeve, M. R., Cosper, T. C., and Walter, M. A. (1975). Visual observations on the process of digestion and the production of faecal pellets in the chaetognath *Sagitta hispida* Conant. *Journal of Experimental Marine Biology and Ecology*, **17**, 39–46.

Rehkämper, G. and Welsch, U. (1985). On the fine structure of the cerebral ganglion of Sagitta (Chaetognatha). *Zoomorphology*, **105**, 83–9.

Reid, J. L., Brinton, E., Fleminger, A., Venrick, E. L., and McGowan, J. A. (1978). Ocean circulation and marine life. In *Advances in oceanography* (ed. H. Charnock and G. Deacon), pp. 65–130. Plenum Press, New York.

Reimer, L. W., Hnatiuk, S., and Rochner, J. (1975). Metacercarien in Planktontieren des mittleren Atlantik. *Wissenschaftliche Zeitschrifte der Pädogogischen Hochschule, Gustrow*, **2**, 239–58.

Reisinger, E. (1969). Ultrastrukturforschung und Evolution. *Bericht der physikalisch-medizinischen Gesellschaft zu Würzburg NF*, **77**, 1–43.

Ricker, W. E. (1975). Computation and interpretation of biological statistics of fish populations. *Bulletin Fisheries Research Board Canada*, **91**, 1–382.

Ritter-Záhony, R. von (1910). Die Chaetognathen. *Fauna Arctica*, **5**, 250–88.

Ritter-Záhony, R. von (1911*a*). Revision der Chaetognathen. *Deutsche Südpolar Expedition 1901–3*, **12**, Hft. 1, 1–72.

Ritter-Záhony, R. von (1911*b*). Chaetognathi. *Das Tierreich (Berlin)*. **29**, 1–35.

Rothschild, M. (1962). Further observations on the growth and trematode parasites of *Peringia ulvae* (Pennant, 1777). *Novitates Zoologicae*, **41**, 84–102.

Rózanska, Z. (1971). Studia nad biologia i ekologia Chaetognatha u Batiku. (Study of the biology and ecology of Chaetognatha in the Baltic). *Dzial Wydavnictw Olsztyn*, 1–88.

Russell, F. S. (1931). The vertical distribution of marine macroplankton. X. Notes on the behaviour of *Sagitta* in the Plymouth area. *Journal of the Marine Biological Association of the United Kingdom*, **17**, 391–407.

Russell, F. S. (1932*a*). On the biology of *Sagitta*. The breeding and growth of *Sagitta elegans* Verrill in the Plymouth area. *Journal of the Marine Biological Association of the United Kingdom*, **18**, 131–46.

Russell, F. S. (1932*b*). On the biology of *Sagitta*. II. The breeding and growth of *Sagitta setosa* J. Müller in the Plymouth area, 1930–1931, with a comparison with that of *Sagitta elegans* Verrill. *Journal of the Marine Biological Association of the United Kingdom*, **18**, 147–60.

Salonen, K. and Sarvala, J. (1980). The effects of different preservative methods on the carbon content of *Megacyclops gigas*. *Hydrobiologia*, **72**, 281–5.

Salvini-Plawen, L. von (1986). Systematic notes on *Spadella* and on the Chaetognatha in general. *Zeitschrift für zoologische Systematik und Evolutionsforschung*, **24**, 122–8.

Salvini-Plawen, L. von (1988). The epineural (vs. gastroneural) cerebral-complex of Chaetognatha. *Zeitschrift für zoologische Systematik und Evolutionsforschung*, **26**, 425–9.

Sameoto, D. D. (1971). Life history, ecological production, and an empirical mathematical model of the population of *Sagitta elegans* in St. Margaret's Bay, Nova Scotia. *Journal of the Fisheries Research Board of Canada*, **28**, 971–85.

Sameoto, D. D. (1972). Yearly respiration rate and estimated energy budget for *Sagitta elegans*. *Journal of the Fisheries Research Board of Canada*, **29**, 987–96.

Sameoto, D. D. (1973). Annual life cycle and production of the chaetognath *Sagitta elegans* in Bedford Basin, Nova Scotia. *Journal of the Fisheries Research Board of Canada*, **30**, 333–44.

Sameoto, D. D. (1987). Vertical distribution and ecological significance of chaetognaths in the arctic environment of Baffin Bay. *Polar Biology*, **7**, 317–28.

Sands, N. J. (1980). Ecological studies on the deepwater community of Korsfjorden, western Norway. *Sarsia*, **65**, 1–12.

Sanzo, L. (1937). Colonia pelagica di uova di Chetognati (*Spadella draco* Krohn). Memori Rendi Compti talassografica Italiano, **239**, 3–6.

Savineau, J-P. and Duvert, M. (1986). Physiological and cytochemical studies of Ca in the primary muscle of the trunk of *Sagitta setosa* (Chaetognath). *Tissue and Cell*, **18**, 953–66.

Schalk, P. H. (1988). Monsoon influences on biogeography and ecology of zooplankton and mikronekton of the Indo-Malayan region, Ph.D. Thesis, University of Amsterdam, pp. 144.

Scharrer, E. (1965). The fine structure of the retrocerebral organ of *Sagitta* (Chaetognatha). *Life Science*, **4**, 923–6.

Schmidt, J. W. (1951). Polarisationoptische untersuchungen an *Sagitta setosa* und *Sagitta hexaptera*.

Zeitschrift für Zellforschung und mikroskopische Anatomie, **35**, 476–85.

Schmidt, J. W. (1951–52). Polarisationoptische untersuchungen an *Sagitta setosa* und *Sagitta hexaptera*. *Zeitschrift für Zellforschung und mikroskopische Anatomie*, **36**, 552–5.

Schram, F. R. (1973). Pseudocoelomates and a nemertine from the Illinois Pennsylvanian. *Journal of Palaeontology*, **47**, 985–9.

Schwartz, L. M. and Stühmer, H. (1984). Voltage dependent sodium channels in an invertebrate striated muscle. *Science*, **225**, 523–5.

Sheader, M. and Evans, F. (1975). Feeding and gut structure of *Parathemisto gaudichaudi* (Guerin) (Amphipoda, Hyperidea). *Journal of the Marine Biological Association of the United Kingdom*, **55**, 641–56.

Shemeleva, A. A. (1965). Weight characteristics of the zooplankton of the Adriatic Sea. *Bulletin de l'Institut océanographique de Monaco*, **65**, pp. 24.

Sherman, K. and Shaner, E. G. (1968). Observations on the distribution and breeding of *Sagitta elegans* (Chaetognatha) in coastal waters of the Gulf of Maine. *Limnology and Oceanography*, **13**, 618–25.

Sheumack, D. D., Howden, M. E. H., Spence, I., and Quinn, R. J. (1978). Maculotoxin: a neurotoxin from the venom glands of the octopus *Hapalochlaena maculosa* identified as tetrodotoxin. *Science*, **199**, 188–9.

Shimazu, T. (1978). Some helminth parasites of the Chaetognatha from Suruga Bay, Central Japan. *Bulletin of the National Science Museum. Tokyo*, ser.A (Zool), **4**, 105–16.

Shimazu, T. (1979). Some protozoan parasites of the Chaetognatha from Suruga Bay, Central Japan. *Japanese Journal of Parasitology*, **28**, 51–5.

Shimizu. Y. (1986). Chemistry and biochemistry of saxitoxin analogues and tetrodotoxin. *Annals of the New York Academy of Science*, **479**, 24–31.

Shipley, A. E. (1901). Chaetognatha. In *Cambridge Natural History*, **2**, pp. 186–94. McMillan, London.

Silas, E. G. and Srinivasan, M. (1968). A new species of *Eukrohnia* from the Indian seas with notes on three other species of Chaetognatha. *Journal of the Marine Biological Association of India*, **10**, 1–33.

Simidu, U., Noguchi, T., Hwang, D. F., Shida, Y., and Hashimoto, K. (1987). Marine bacteria which produce tetrodotoxin. *Applied Environmental Microbiology*, **53**, 1714–5.

Slabber, M. (1778). Natuurkundige Verlustigingen behelzende microscopise Waarneemingen van in-en uitlandse Water-en Land-dieren, p. 166. J. Bosch, Haarlem.

Sochard, M. R., Wilson, D. F., Austin, B., and Colwell, R. R. (1979). Bacteria associated with the surface and gut of marine copepods. *Applied Environmental Marine Biology*, **37**, 705–59.

Spero, H., Hagan, D., and Vastano, A. (1979). An SEM examination of *Sagitta tenuis* (Chaetognatha) using a special sedation and handling procedure. *Transactions of the American Microscopical Society*, **98**, 139–41.

Srinivasan, M. (1986). *Pterokrohnia arabica*, a new genus and species of Chaetognatha from the Arabian Sea. *Journal of the Marine Biological Association of India*, **28**, 199–201.

Stadel, O. (1958). Die Chaetognathen-Ausbeute. *Deutsche Antarktische Expedition*, **1935/39**, 208–43.

Steedman, H. F. (1976*a*). Osmotic pressures in fixation and preservation. In *Zooplankton fixation and preservation*. Monographs on oceanographical methodology, **4**, (ed. H. F. Steedman,) pp. 186–8. Unesco Press, Paris.

Steedman, H. F. (1976*b*). General and applied data on formaldehyde fixation and preservation of marine zooplankton. In *Zooplankton fixation and preservation*, Monographs on oceanographic methodology, **4**, (ed. H. F. Steedman,) pp. 103–54. Unesco Press, Paris.

Steinhaus, O. (1896). Die Verbreitung der Chaetognathen im südatlantischen Ozean, Thesis, Christan-Albrecht Universität, Kiel, pp. 49.

Stevens, N. M. (1903). On the ovogenesis and spermatogenesis of *Sagitta bipunctata*. *Zoologische Jahrbücher für Anatomie*, **18**, 227–40.

Stevens, N. M. (1910). Further studies on reproduction in *Sagitta*. *Journal of Morphology*, **21**, 279–303.

Stone, J. H. (1966). The distribution and fecundity of *Sagitta enflata* Grassi in the Agulhas Current. *Journal of Animal Biology*, **35**, 533–41.

Stone, J. H. (1969). The Chaetognatha community of the Agulhas Current: its structure and related properties. *Ecological Monographs*, **39**, 433–63.

Strathmann, M. F. and Shinn, G. L. (1987). Phylum Chaetognatha. In *Reproduction and development of marine invertebrates of the northern Pacific coast*. (ed. M. F. Strathmann), pp. 647–56. University of Washington Press, Seattle.

Suarez-Caabro, J. A. (1955). Quetognatos de los mares cubanos. *Memorias de la Sociedad Cubana de Historia Natural*, **22**, 125–89.

Sullivan, B. K. (1977). Vertical distribution and feeding of two species of chaetognaths at Weather Station P, PhD Thesis, University of Hawaii, p. 147.

Sullivan, B. K. (1980). *In situ* feeding behaviour of *Sagitta elegans* and *Eukrohnia hamata* (Chaetognatha) in relation to the vertical distribution and abundance of prey at ocean station 'P'. *Limnology and Oceanography*, **25**, 317–26.

Sund, P. (1961). Two new species of Chaetognatha from the waters off Peru. *Pacific Science*. **15**, 105–11.

Sweatt, A. J. (1980). Chaetognaths in lower Narragansett Bay. *Estuaries*, **3**, 106–10.

Sweatt, A. J. and Forward, R. B. (1985a). Diel vertical migration and photoresponses of the chaetognath *Sagitta hispida* Conant. *Biological Bulletin. Marine Biological Laboratory, Woods Hole*, **168**, 18–31.

Sweatt, A. J. and Forward, R. B. (1985b). Spectral sensitivity of the Chaetognath *Sagitta hispida* Conant. *Biological Bulletin. Marine Biological Laboratory, Woods Hole*, **168**, 32–8.

Szaniawski, H. (1983). Chaetognath grasping spines recognised among Cambrian protoconodonts. *Journal of Palaeontology*, **56**, 806–10.

Szyper, J. P. (1976). The role of *Sagitta enflata* in the southern Kaneohe Bay system, PhD Thesis, University of Hawaii, p. 147.

Szyper, J. P. (1978). Feeding rate of the chaetognath *Sagitta enflata* in nature. *Estuarine and Coastal Marine Science*, **7**, 567–75.

Tamplin, M. L., Colwell, R. R., Hall, S., Kogure, K., and Strichartz, G. R. (1987). Sodium channel inhibitors produced by enteropathogenic *Vibrio cholerae* and *Aeromonas hydrophila*. *Lancet*, **i**, 975.

Tande, K. S. (1983). Ecological investigations of the zooplankton community of Balsfjorden, northern Norway: Population structure and breeding biology of the chaetognath *Sagitta elegans* Verrill. *Journal of Experimental Marine Biology and Ecology*, **68**, 13–24.

Taw, N. (1974). A new species of *Sagitta* (Chaetognatha) from d'Entrecasteaux Channel, Tasmania. *Papers and Proceedings of the Royal Society of Tasmania*, **109**, 77–81.

Tchindonova, J. G. (1955). Chaetognatha of the Kurile-Kamchatka Trench. *Trudy Instytutu Okeanology*, **12**, 298–310.

Terazaki, M. and Marumo, R. (1982). Feeding habits of meso- and bathypelagic chaetognatha. *Sagitta zetesios* Fowler. *Oceanologica Acta*, **5**, 461–4.

Terazaki, M. and Miller, C. B. (1982). Reproduction of meso- and bathypelagic chaetognaths in the genus *Eukrohnia*. *Marine Biology*, **71**, 193–6.

Terazaki, M. and Miller, C. B. (1986). Life history and vertical distribution of pelagic chaetognaths at Ocean Station P in the subarctic Pacific. *Deep-Sea Research*, **33**, 323–37.

Terazaki, M., Marumo, R., and Fujita, Y. (1977). Pigments of meso- and bathypelagic chaetognaths. *Marine Biology*, **41**, 119–25.

Thomson, J. (1947). The Chaetognatha of south-eastern Australia. *Bulletin. Council for Scientific and Industrial Research. Melbourne*, **222**, 1–43.

Thuesen, E. V. and Bieri, R. (1987). Tooth structure and buccal pores in the chaetognath *Flaccisagitta hexaptera* and their relation to the capture of fish larvae and copepods. *Canadian Journal of Zoology*, **65**, 181–7.

Thuesen, E. V. and Kogure, K. (1989). Bacterial production of tetrodotoxin in four species of chaetognaths. *Biological Bulletin*, **176**, 191–4. *Marine Biological Laboratory, Woods Hole*.

Thuesen, E. V., Kogue, K., Hashimoto, K., and Nemoto, T. (1988a). Poison arrowworms: a tetrodotoxin venom in the marine phylum Chaetognatha. *Journal of Experimental Marine Biology and Ecology*, **116**, 249–56.

Thuesen, E. V., Nagasawa, S., Bieri, R., and Nemoto, T. (1988b). Transvestibular pores of chaetognaths with comments on the function and nomenclature of the vestibular anatomy. *Bulletin of Plankton Society of Japan*, **35**, 133–41.

Tiselius, P. T. and Peterson, W. T. (1986). Life history and population dynamics of the chaetognath *Sagitta elegans* in central Long Island Sound. *Journal of Plankton Research*, **8**, 183–95.

Tokioka, T. (1938). A new chaetognath (*Sagitta crassa* n. sp.) from Ise Bay. *Zoological Magazine. Tokyo*, **50**, 349–51.

Tokioka, T. (1939). Chaetognaths collected chiefly from the bays of Sagami and Suruga, with some notes on the shape and structure of the seminal vesicle. *Records Oceanographic Works Japan*, **10**, 122–50.

Tokioka. T. (1940). The chaetognath fauna of the waters of western Japan. *Records Oceanographic Works Japan*, **10**, 1–22.

Tokioka, T. (1942). Systematic studies of the plankton organisms occurring in Iwayama Bay, Palao. III. Chaetognaths from the Bay and adjacent waters. *Contributions of the Seto Marine Biological Laboratory*, **104**, 527–48.

Tokioka, T. (1950). Notes on the development of the eye and the vertical distribution of chaetognatha. *Contribution Paper of the Seto Marine Biological Laboratory*, **132**, 117–32.

Tokioka, T. (1952). Chaetognaths of the Indo-Pacific. *Annotationes Zoologicae Japonenses*, **25**, 307–16.

Tokioka. T. (1959). Observations on the taxonomy and distribution of chaetognaths of the North Pacific. *Publications of the Seto Marine Biological Laboratory*, **7**, 349–456.

Tokioka, T. (1962). The outline of the investigations made on chaetognaths in the Indian Ocean. *Information Bulletin of Planktology*. *Japan*, **8**, 5–11.

Tokioka, T. (1965a). The taxonomical outline of chaetognaths. *Publications of the Seto Marine Biological Laboratory*, **12**, 335–57.

Tokioka, T. (1965b) Supplementary notes on the systematics of Chaetognatha. *Publications of the Seto Marine Biological Laboratory*, **13**, 231–42.

Tokioka, T. (1965c). Chaetognatha. In *Animal phylogenetic systematics*. (ed. T. Uchida) vol. 8, pp. 259–92. Nakayama Bookstore Ltd, Tokyo.

Tokioka, T. (1974a). Morphological differences observed between the generations of the same chaetognath population. *Publications of the Seto Marine Biological Laboratory*, **21**, 269–79.

Tokioka, T. (1974b). On the specific validity of species pairs or trios of plankton animals distributed respectively in different but adjoining water masses as seen in chaetognaths. *Publications of the Seto Marine Biological Laboratory*, **21**, 393–408.

Tokioka, T. (1979). Neritic and oceanic plankton. In *Zoogeography and diversity of plankton* (ed. S. Van der Spoel and A. C. Pierrot-Bults), 126–43. Bunge, Utrecht.

Tokioka, T. and Pathansali, D. (1964). *Spadella cephaloptera forma angulata* raised to the rank of species. *Publications of the Seto Marine Biological Laboratory*, **12**, 145.

Tokioka, T. and Pathansali, D. (1965). A new form of *Sagitta bedoti* Beraneck found in the littoral waters near Penang. *Bulletin of the National Museum, Singapore*, **33**, 1–5.

Trégouboff, G. (1949). Un parasite nouveau des sagittes (note préliminaire). *Bulletin de l'Institut oceanografique de Monaco*, **953**, 1–6.

Tuzet, O. (1931). Recherches sur la spermatogenèse des chaetognathes *Sagitta bipunctata* (Quoy et Gaim) et *Spadella cephaloptera* (Busch). *Archives Zoologie Experimentale et Générale*, **71**, 1–15.

Ussing, H. H. (1938). The biology of some important plankton animals in the fjords of east Greenland. *Meddedelser om Grönland*, **100**, 1–108.

Uye, S. (1982). Length-weight relationships of important zooplankton from the Inland Sea of Japan. *Journal of the Oceanographical Society of Japan*, **38**, 149–58.

Van der Spoel, S. (1971). Some problems in infraspecific classification of holoplanktonic animals. *Zeitschrift für zoologische Systematik und Evolutionsforschung*, **9**, 107–38.

Van der Spoel, S. (1983). Patterns in plankton distribution and their relation to speciation. The dawn of pelagic biogeography. In *Evolution, time and space: the emergence of the biosphere* (ed. R. W. Sims, J. H. Price, and P. E. S. Whalley), pp. 291–334. Academic Press, London.

Van der Spoel, S. and Pierrot-Bults, A. C. (ed.) (1979a). *Zoogeography and diversity of plankton*, pp. 1–410. Bunge, Utrecht.

Van der Spoel, S. and Pierrot-Bults, A. C. (1979b). Zoogeography of the Pacific Ocean. In *Zoogeography and diversity of plankton* (ed. S. van der Spoel and A. C. Pierrot-Bults), pp. 291–327. Bunge, Utrecht.

Van der Spoel, S. and Heyman, R. P. (1983). *A comparative atlas of zooplankton*, pp. 1–186. Bunge, Utrecht.

Van der Spoel, S. and Schalk, P. H. (1988). Unique deviations in depth distribution of the deep-sea fauna. *Deep-Sea Research*, **35**, 1185–93.

Vanucci, M. and Hosoe, K. (1952). Resultados científicos do cruzeiro do 'Baependi' e do 'Vega' à Ilha da Trindade. Chaetognatha. *Boletim do Instituto oceangrafico. São Paulo*, **3**, 5–30.

Vasiljev, A. (1925). La fécondation chez *Spadella cephaloptera* Lgrhs. et l'origine du corps déterminant la voie germinative. *Biologie Générale*, **1**, 249–78.

Vinogradov, M. E. (1970). *Vertical migration of the oceanic plankton*, pp. 1–339. Israel Program for Scientific Translations, Jerusalem.

Wainwright, S. A. (1982). Structural systems: hydrostats and frame works. In *A companion to animal physiology* (ed. E. R. Taylor, K. Johansen, and L. Bobis). Cambridge University Press.

Walford, L. A. (1958). *Living resources of the sea*, p. 321. Ronald Press, New York.

Weinstein, M. (1972). Studies on the relationship between *Sagitta elegans* Verrill and its endoparasites in the southwestern Gulf of St. Lawrence, PhD Thesis, McGill University, Montreal, p. 202.

Welsch, U. and Storch, V. (1982). Fine structure of

the coelomic epithelium of *Sagitta elegans* (Chaetognatha). *Zoomorphology*, **100**, 217–22.

Welsh, U. and Storch, V. (1983*a*). Enzymhistochemische und elektronenmikroskopische Beobachtungen am Darmepithel von *Sagitta elegans*. *Zoologische Jahrbücher, Abteilung für Anatomie*, **109**, 23–33.

Welsch, U. and Storch, V. (1983*b*). Fine structural and enzyme histochemical observations on the epidermis and the sensory cells of *Sagitta elegans* (Chaetognatha). *Zoologischer Anzeiger*, **210**, 34–43.

White, B. N. (1987). Oceanic anoxic events and allopatric speciation in the deep sea. *Biological Oceanography*, **5**, 243–59.

Williams, R. and Robins, D. B. (1982). Effects of preservation on wet weight, dry weight, nitrogen and carbon contents of *Calanus helgolandicus* (Crustacea, Copepoda). *Marine Biology*, **71**, 271–81.

Wimpenny, R. S. (1937). The distribution, breeding and feeding of some important plankton organisms of the south-west North Sea in 1934. I. *Calanus finmarchicus* (Gunn), *Sagitta setosa* (J. Müller), and *Sagitta elegans* (Verrill). *Fishery Investigations London*, Series 2, **15**, 1–53.

Wimpenny, R. S. (1938). Diurnal variation in the feeding and breeding of zooplankton related to the numerical balance of the zoo-phytoplankton community.

Journal du Conseil Permanente International pour l'Exploration de la Mer, **13**, 323–37.

Yamaji, I. (1980). *Illustrations of the marine plankton of Japan*, pp. 262–78. Hoikusha Publishing Company, Osaka.

Yotsu, M., Yamakazi, Y., Meguro, Y., Endo, A., Murata, M., Naoki, H., and Yasumoto, T. (1987). Production of tetrodotoxin and its derivatives by *Pseudomonas* sp. isolated from the skin of a pufferfish. *Toxicon*, **25**, 225–8.

Zaika. V. E. (1972). Growth and specific production of *Sagitta setosa* in the Black Sea. In *Specific production of aquatic invertebrates*, pp. 97–100. Israel Program for Scientific Translations (1973). Wiley, New York.

Zhang, G. and Chen, Q-C. (1983). Studies on chaetognaths in the central and northern parts of the South China Sea. *Contributions on marine biological Research of the South China Sea*, **1**, 17–63 (in Chinese).

Zo, Z. (1973). Breeding and growth of the chaetognath *Sagitta elegans* in Bedford Basin. *Limnology and Oceanography*, **18**, 750–6.

ZoBell, C. E. (1972). Substratum-bacteria, fungi and blue-green algae. In *Marine Ecology*, 1 (ed. O. Kinne), pp. 1251–70. Wiley-Interscience, New York.

TAXONOMIC INDEX

SUBJECT INDEX